ROUTING

PROJECTS & TECHNIQUES FOR THE SMALL ROUTER

ROUTING

PROJECTS & TECHNIQUES FOR THE SMALL ROUTER

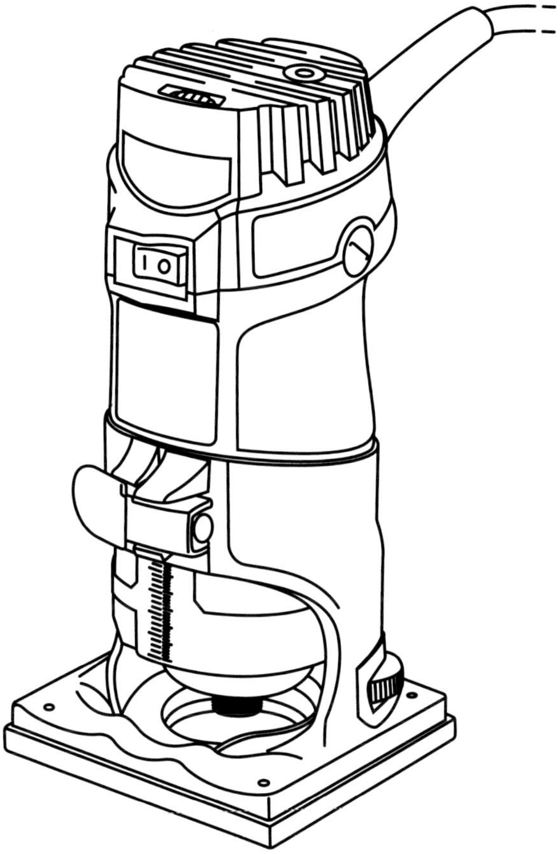

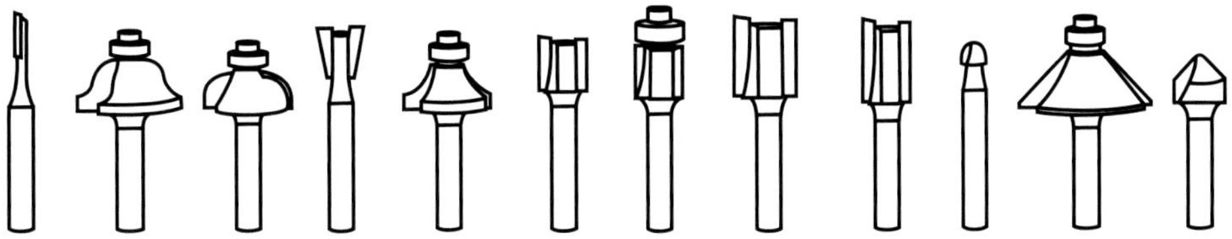

TOYOHISA SUGITA

Contents

Tips

A Magical Tool for Opening Your Woodworking Future

COMPLETELY NEW TECHNIQUES FOR THE 'TRIMMER'

As many woodworking enthusiasts may know, a 'trimmer' is a woodworking power tool that cuts material with a rotating bit. It is primarily used for chamfering edges or adding decorative details to wood. However, some may find that after acquiring one, they end up using it less frequently than expected and may even feel that it's just sitting around the workshop. That would be a real waste.

Making traditional joints like 'mitre joints' or 'box joints' once required specialist skills and a sharp eye. However, with a trimmer, even such precise work can be performed by almost anyone. The book you are holding in your hands is chock-full of techniques that will unlock the hidden potential of trimmers and allow anyone to engage in high-quality woodworking, without the need for advanced skills.

Learn entirely new ways of using a trimmer that will elevate your woodworking hobby to the level of a professional. And, more importantly, enjoy mastering the trimmer along the way.

WHAT YOU WILL LEARN IN THIS BOOK

In Chapter 1, we explain 'Trimmer Basics', followed by 'Handheld Operations' in Chapter 2, and how to use a 'Trimmer Table' in Chapter 3. Building on this knowledge, Chapter 4, 'Projects That Make Full Use of a Trimmer', introduces innovative processing methods through various sample projects. Each project incorporates important techniques, so it's a good idea to read the explanations below and start with the ones that catch your eye.

EASY TO MAKE MITRE JOINT BOX 1

P.58

Those who are well-versed in woodworking are likely aware of the difficulty of a 'mitre joint'. Achieving an accurate 45-degree cut significantly raises the bar when engaged in serious woodworking. In this project, we'll introduce an unexpectedly simple method for creating a mitre joint using just a V-groove bit. This project will also allow you to start handling a trimmer.

INSET INRO BOX

P.100

This project involves creating an 'inset inro box', a type of Japanese box where the lid and body fit together seamlessly. The protrusion on the body (where the lid fits) is made from a separate material. That means we need to use mitred joinery again. The trimmer will be used for other processes as well, such as cutting the grooves to fit the bottom panel and separating the lid from the body.

BOX WITH ATTACHED INRO

P.108

The next challenge is to tackle the more advanced 'box with attached inro'. Here, the box material itself is directly modified with a 'rabbet cut' to create the protrusions on the body. This requires a high level of precision, which can be achieved effectively by utilizing a trim table.

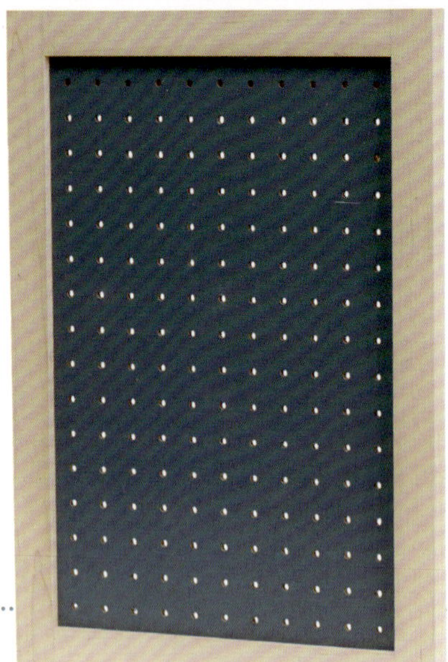

PEGBOARD

P.114 In this guide, you will learn how to create 'cope and stick joints' – useful for making frames and doors – as well as 'mortise' joints, one of the classic joint types. Using a trimmer allows for high-precision groove and rabbet cutting. Additionally, the guide will explain how to make an excellent 'French cleat' for wall mounting.

HALF-BLIND RABBET JOINT BOX

P.124 Once you can easily perform rabbet and groove machining, you can create a 'half-blind rabbet joint'. The lid itself will also have a rabbet to ensure a snug fit when closed.

WALL-MOUNTED SHELF

P.136 This example uses 'dado joints' made with a 'T-square sliding fence' (page 40). These techniques can be applied to the production of other shelves. The guide also covers methods for adjusting the position and width of grooves, as well as advanced uses of the edge guide.

BOX JOINT BOX

P.154 'Box joints', which traditionally required high-level skills with chisels and other hand tools, can be made easily with a trimmer. For this, we will use the 'box joint jig' introduced on page 146. Feel free to refer to that section as needed.

BOX JOINT INRO BOX

P.162 This box uses box joints for the fitted 'inro' portion. Instead of using conventional pins, it makes use of the box joint's interlocking features. The lid and body are originally made from a single piece of wood. This maintains continuous wood grain after processing the box joints and also during their subsequent separation. For the lid attachment, a 'Small Tenon Half-Blind Rabbeted Dado Joint' (page 167) is used.

PAPER TOWEL HOLDER

P.170 This is a simple holder that uses box joints for the back and side panels. The curved sections are shaped using a trimmer and template. Additionally, a unique trimmer method for creating a rounded shoulder joint – by thinning part of the round rod – is introduced.

DESKTOP CHEST

P.180 Despite its compact size, this chest allows you to thoroughly learn how to create well-crafted drawers. The techniques and cutting steps for each part are rich in know-how, making this a rewarding project – even for advanced woodworkers.

PHOTO FRAME

P.198 This is a mitred photo frame. The recesses on the back for holding the photo, along with the decorative parts on the front, are all created with trimmer table rabbet joints. A jig for accurately cutting a 45-degree angle is also introduced. This is truly a simple, yet enjoyable, example of woodworking with hand tools.

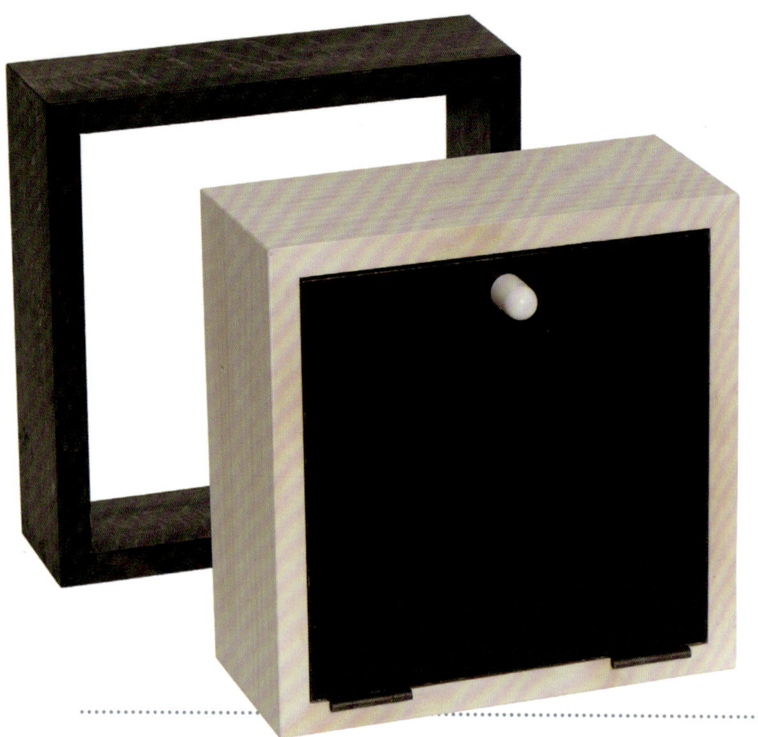

WALL-MOUNTED BOXES

P.206 Here we are introduced to spline joints. They are often used to reinforce large mitre joints by adding a groove and an inserted wooden shim. Various jigs – such as the 45-degree angled jig (page 176) and the edge guide jig (page 45) – must be used, so make sure to prepare them in advance. Additionally, the trimmer is used to create a lightweight flush panel lid.

CUTTING BOARD

P.222 After gluing the material together, you will create a beautifully cut two-tone cutting board by levelling the surface with a trimmer planer (page 36). Edges are rounded and the surrounding grooves are cut using the trimmer and a template.

FEATHERBOARD

P.234 This oddly shaped object is actually a tool! It's designed to safely and securely press wood against the trimmer table using the wood's own natural tension. It can be made using just a saw and a simple jig. That makes it a great practice project to prepare yourself for working with a trimmer table.

1

Chapter 1

Trimmer Basics

① Main Switch
⑦ Trimmer Bit
⑤ Collet Chuck (Nut)
② Hose Clamp
③ Housing
⑧ Clamp
④ Edge-Guide Mounting Screw
⑤ Collet Chuck (Nut)
⑥ Base Plate (Homemade)

Basic Trimmer Operation

First, I will explain the basics of handling a trimmer. Those of you who already have some woodworking experience and are trying to master the trimmer may feel that this is unnecessary. However, a little base knowledge always serves as a foundation for more complex techniques, so as a primer, I recommend giving this section a read.

Trimmer Part Names

① **Main Switch**	Switch for turning trimmer on and off. Make absolutely certain that the trimmer bit isn't touching anything when you turn it on!
② **Hose Clamp**	This was added by the author. It serves as a stopper for the housing. It can be used to fix the cutting depth as well.
③ **Housing**	The housing slides, so be sure to use the lever to lock it in place at the desired cutting depth. The housing also serves as a safety cover for the bit.
④ **Edge-Guide Mounting Screw**	This mounting screw is for attaching the straight guide (edge guide) introduced on page 16.
⑤ **Collet Chuck (Nut)**	This is the nut for attaching the router bit. Instructions for its use can be found on pages 14–15.
⑥ **Base Plate (Homemade)**	This plate stabilizes the trimmer during handheld operation. We introduce a method for making a larger, easier-to-use DIY version (page 32 and following pages).
⑦ **Trimmer Bit**	Different bits enable various types of cutting. We introduce representative bit types on pages 17–19.
⑧ **Clamp**	This part fixes the housing in place. It also fixes the cutting depth.

Inserting and Removing Trimmer Bits

First, unplug the power cord from the outlet. Then, remove the main unit from the transparent housing.

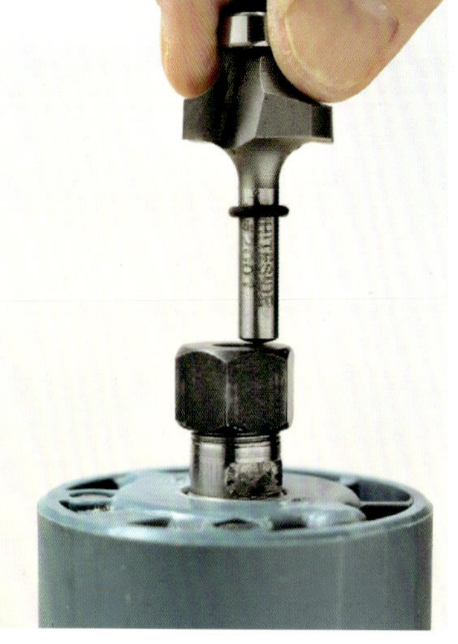

Here we see a trimmer bit after removal. Notice that there is a small rubber O-ring attached to the shaft. This allows us to set the bit at a consistent depth.

When the O-ring is inserted to the indicated position, the lower end of the shaft (shank) will be about 5/64in (2mm) away from the chuck bottom (see collet chuck mechanism on the next page).

If the shaft is inserted too deeply – to the point where the neck begins to spread – it won't be securely fixed. Pay attention as this can be quite dangerous!

TIGHTEN

Tighten the hexagonal collet nut firmly with the two included wrenches (depends on the model, some may only come with one wrench if there is a stopper button on the motor or shaft). You can also tighten with one hand while holding the trimmer, as shown in the photo on the left.

LOOSEN

Loosen the collet nut to remove the trimmer bit. The two wrenches will be opposite when tightening. Be careful not to pinch your fingers if the wrenches come off the nut while loosening or tightening.

◉ COLLET CHUCK MECHANISM

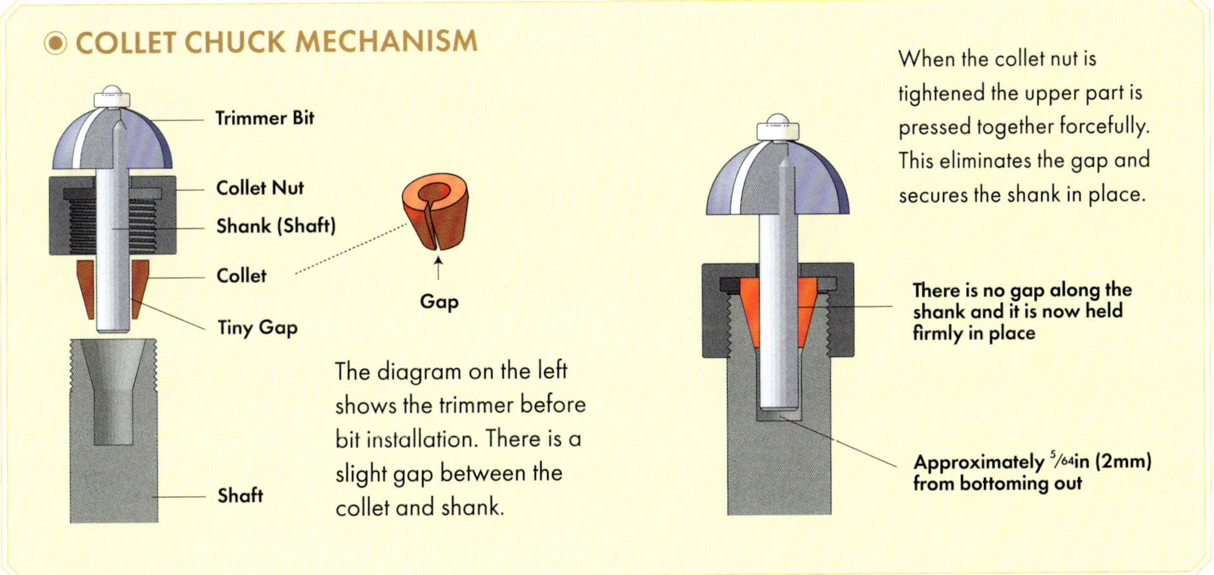

The diagram on the left shows the trimmer before bit installation. There is a slight gap between the collet and shank.

When the collet nut is tightened the upper part is pressed together forcefully. This eliminates the gap and secures the shank in place.

◉ INCH-SIZED COLLETS

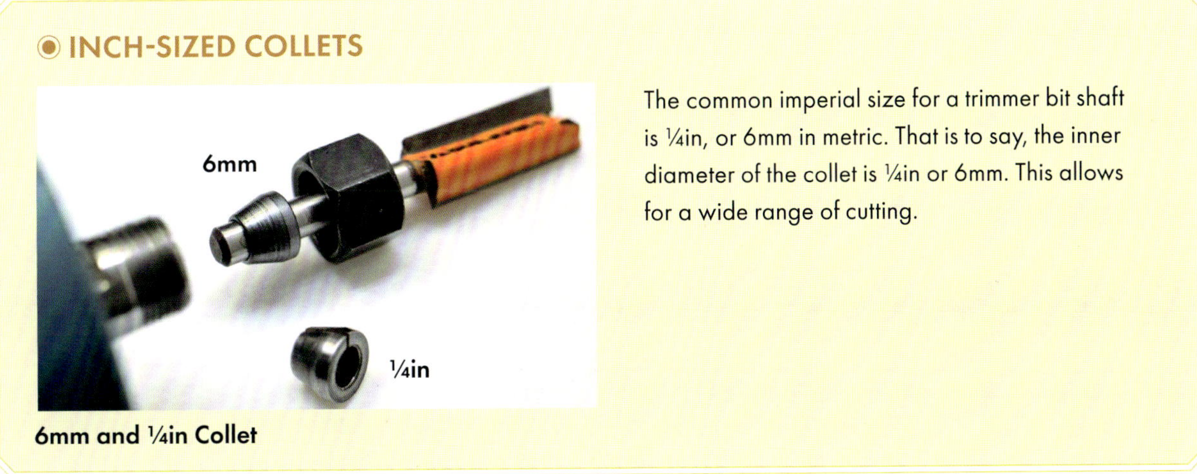

6mm and ¼in Collet

The common imperial size for a trimmer bit shaft is ¼in, or 6mm in metric. That is to say, the inner diameter of the collet is ¼in or 6mm. This allows for a wide range of cutting.

Setting the Trimmer Bit Depth

Once the trimmer bit is attached, we set the cutting depth. As shown in the photo, place a board with the same thickness as the desired depth (a spacer) on a flat surface (wood or acrylic board), and then set the trimmer on top. In this case, a ⅜in (10mm) transparent board spacer is set on a thin piece of wood.

Next, lower the trimmer bit until it touches the base plate and lock it in place. This sets the cutting depth. Note that you can also make your own spacers out of plywood or MDF by drilling holes or channels in them.

In the photo, a hose clamp (steel belt used primarily to secure a hose to a water faucet) is attached to the trimmer body. In this case, it's used as a depth stop. Note that the hose clamp was fixed so that it touches the transparent base of the trimmer with the maximum depth set to ⅜in (10mm).

How to Hold a Trimmer

When holding the trimmer, make sure to grip OPPOSITE the opening of the transparent housing (as shown in the photo) rather than the opening itself. This will prevent your fingers from entering the housing. Additionally, you should note that the power cord often gets in the way during operation. Many people work with it draped over their shoulder. Find a method that works best for you.

Straight Guide (Edge Guide)

A straight guide, or edge guide, is attached to the trimmer and is used by running it along the material's edge. It creates grooves, or rabbet cuts, that are parallel to any edge. In the photo below left, I attached a square rod, using wood ¾×1×11in (18×25×280mm) in size. This provides more stability during operation. In the right-hand photo, you can see that a groove cut has been made on the near side of the material.

Place the straight guide against the near edge of the material and cut the groove.

Various Types of Trimmer Bits

Trimmer bits come in a variety of shapes. Use them according to your needs. The more types of bits you have, the wider the range of operations you can perform.

① SPIRAL BITS

The photo on the left shows an up-cut spiral bit. The one on the right is a down-cut spiral bit. The blades on these bits are spiral in shape and are always in contact with the material. This results in a cleaner cut when compared to straight bits. In particular, down-cut spiral bits reduce edge splintering when cutting grooves.

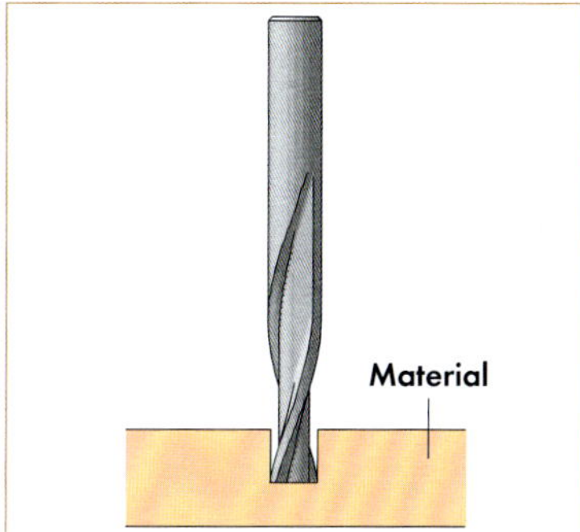

◉ SPIRAL BIT COMPARISON

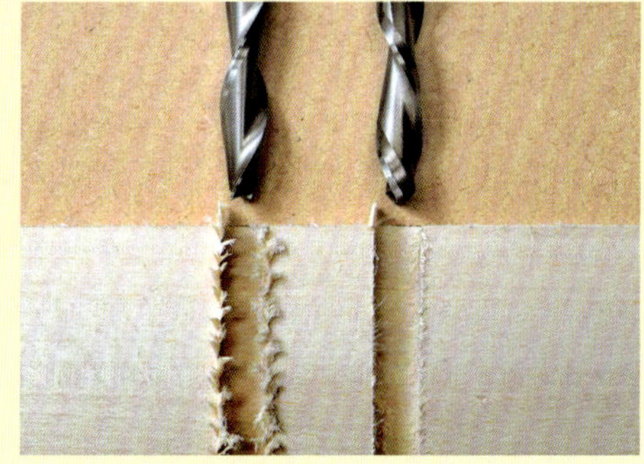

Even though spiral bit rotation directions are the same, the twist on the blades is reversed. This results in different finish cuts. When performing handheld groove cutting operations, the down-cut spiral bit provides a cleaner finish. Conversely, when using a down-cut spiral bit on a trimmer table, the force applied tends to lift the material, which can be dangerous. In this case, the up-cut spiral bit – which pulls the material downwards into the tabletop – is more suitable.

Left: Up-cut Spiral Bit, Right: Down-cut Spiral Bit

② STRAIGHT BITS

There are various types – and names – for straight bits, based on differences in blade diameter and length. When planing wide and flat material, larger diameter bits are used, while thinner bits are more suitable for cutting grooves.

Manufacturers: From Left
EAGLE AMERICA, PRC, CMT

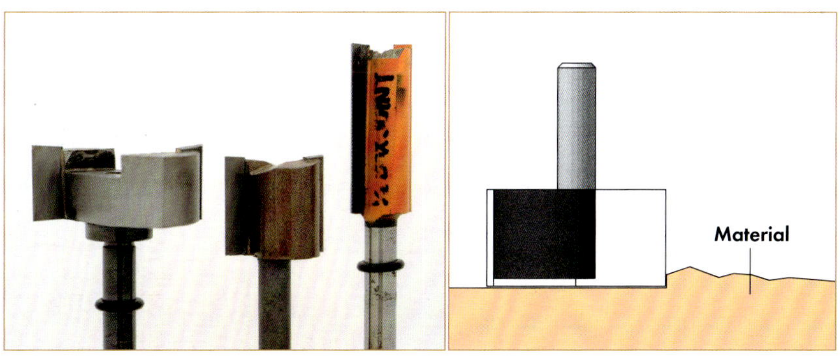

③ PATTERN BITS & FLUSH TRIM BITS

Pattern bits have bearings at their base and are used to shape material to match a template (see page 23). Flush trim bits have a bearing at the tip and are used to process panels made from veneered plywood, or similar materials, glued to a wooden frame (see page 218).
Manufacturer: MLCS

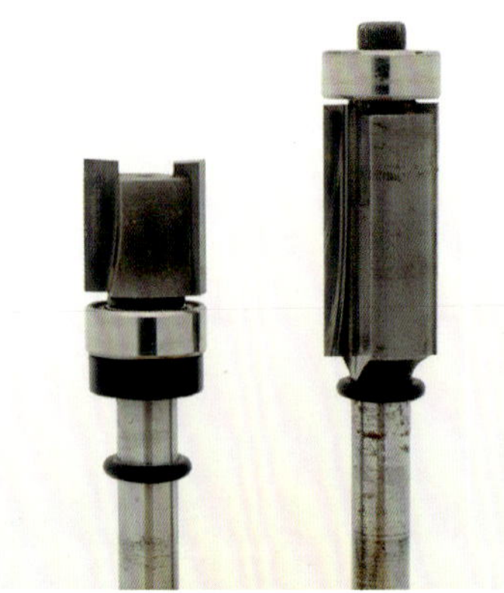

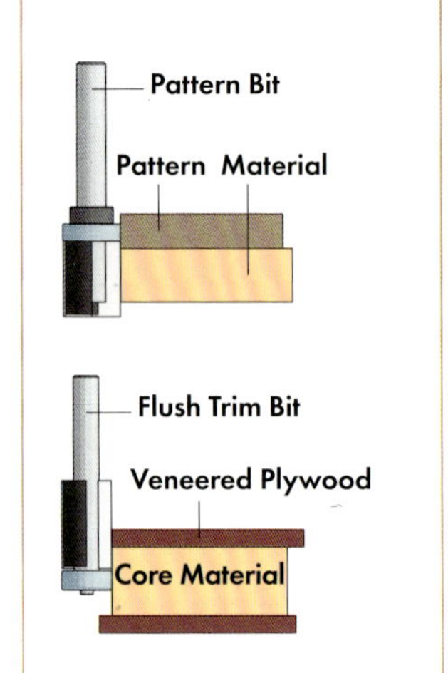

④ SPECIAL PATTERN BITS

These bits tend to have extremely short blade lengths. They have bearings, which makes them suitable for dado joints and hinge mortising. The illustration on the right shows the process of creating a dado groove between two templates.
Manufacturer: MLCS

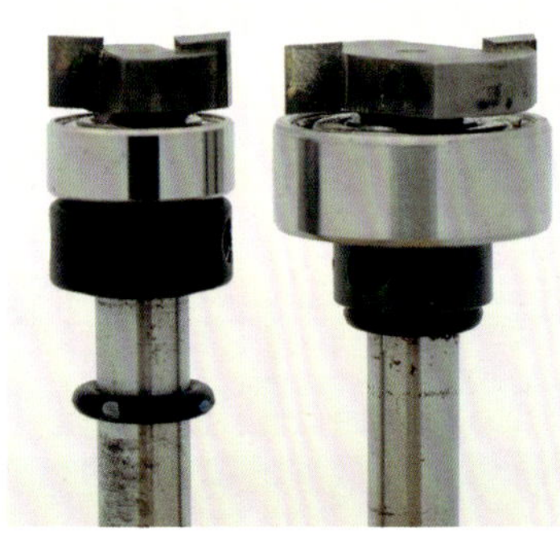

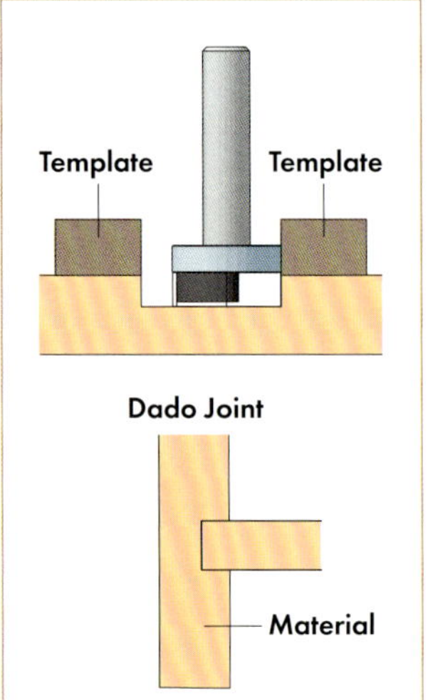

⑤ CHAMFER BIT/ROUND-OVER BIT

The chamfer bit on the left creates a 45-degree bevel by cutting while the bearing is against the material. The length of the cutting surface determines the size of the bevel. The round-over bit on the right smoothly curves edge material. There are various other shapes available.
Manufacturer: WHITESIDE

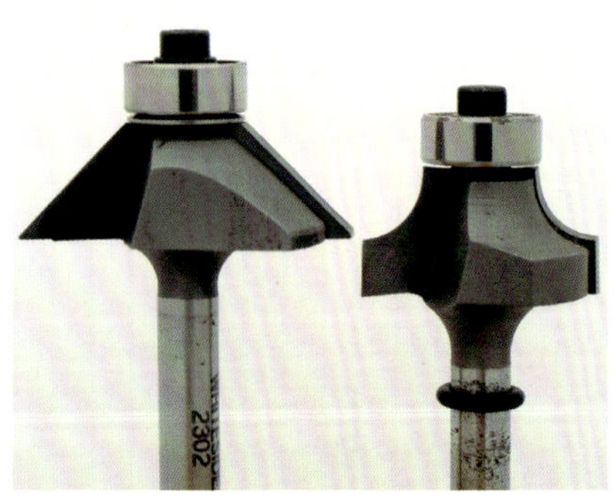

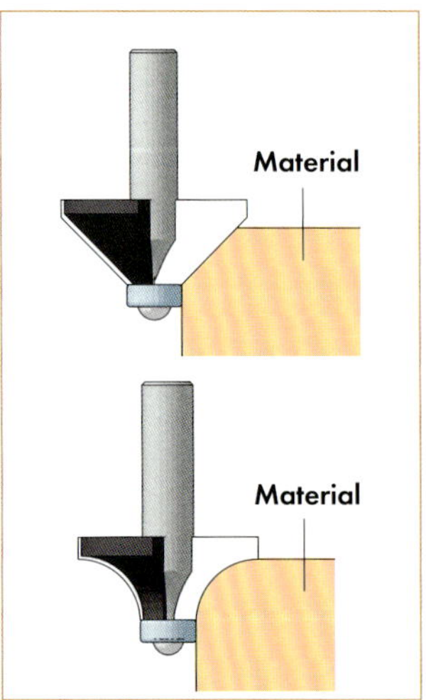

⑥ DOVETAIL BITS

These bits are used for dovetail joints. There are various bit angles – such as 14-degrees or 8-degrees. Choose your angle based on the intended purpose.

Manufacturer: From Left
CMT, EAGLE AMERICA

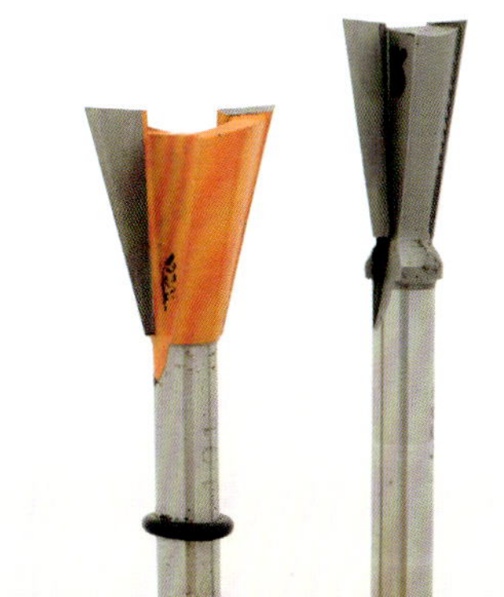

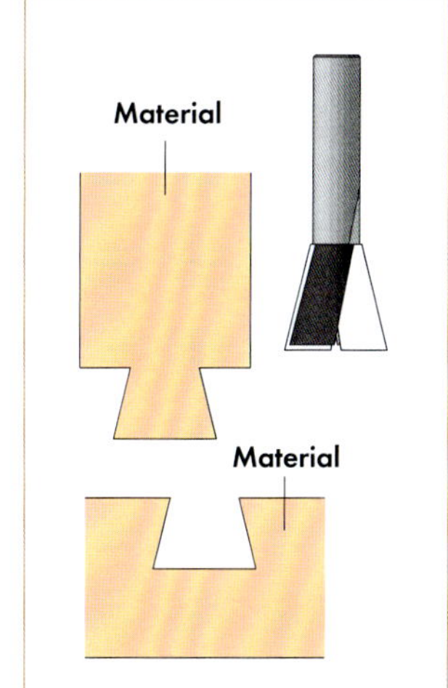

⑦ DISH CUTTER

Dish cutters are suitable for making items like coasters because of their rounded corners. In this book, we also use them when finishing thin boards to cover plywood edges.

Seller: ROCKLER

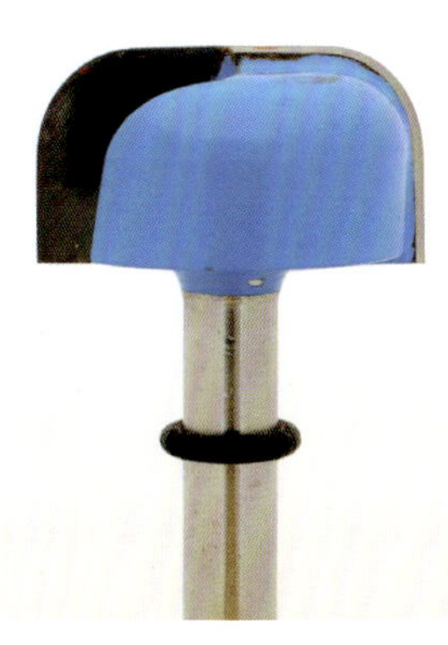

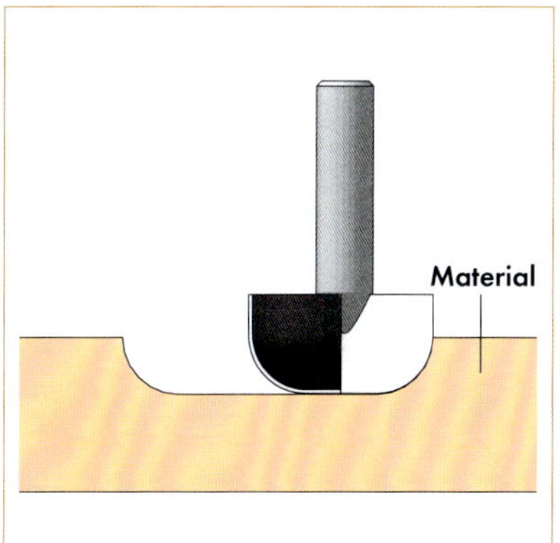

⑧ U-SHAPED & V-SHAPED BITS

The two U-shaped bits on the left of the photo are used for grooving cutting boards (see page 230). The two V-shaped bits on the right have an angle of 90-degrees. They are convenient for joinery, and in this book, they are used to make V-grooves that can form boxes (see pages 59, 103 and 110).

Manufacturers: From Left WHITESIDE, WHITESIDE, Justool, MLCS (router bit)

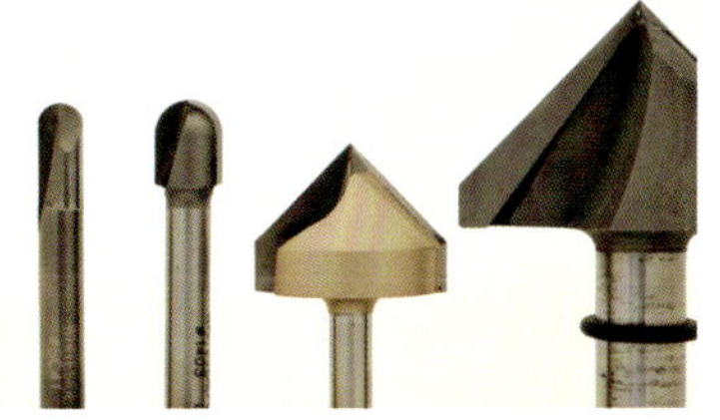

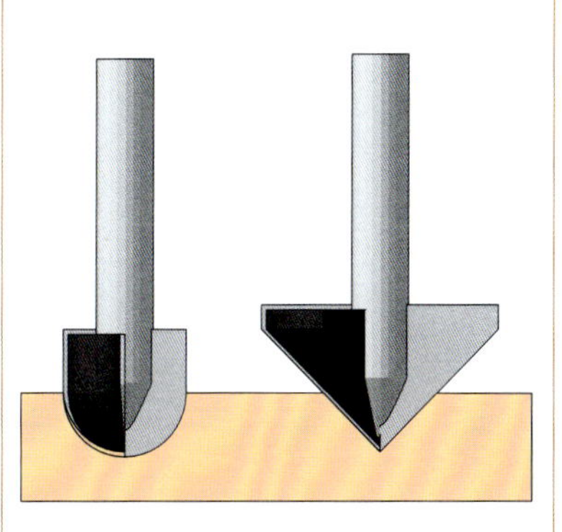

The Tendency to Curve Left

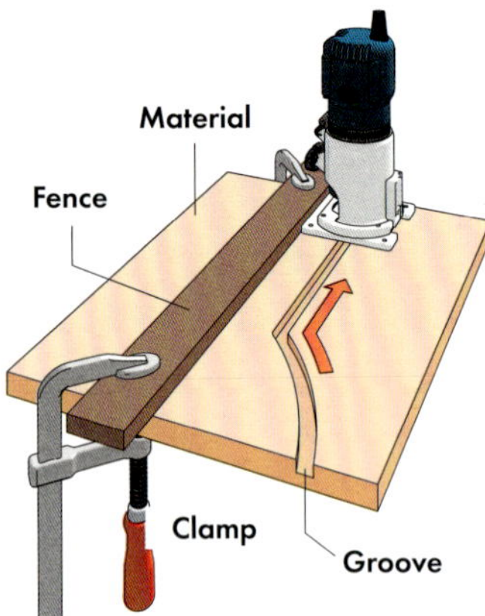

When using a trimmer for groove cutting, it always has a tendency to pull to the left (based on direction of travel). To take advantage of this tendency when making straight grooves, a straight edge called a fence is used. Any straight object can serve as a fence. The fence is attached to the left of the trimmer's direction of travel.

Diagram A on the right shows a top-down view of a bit attached to a trimmer. The diagram illustrates the process of groove cutting, to the bit's diameter, as indicated by the dotted line. We have a fence (straight edge) clamping the material down on the left side. When the trimmer is pushed along for groove cutting, the bit rotates clockwise. This causes the trimmer to be influenced by the rotation of the bit, and the trimmer will move to the left as indicated by the red arrow. If you think of the tire depicted next to the trimmer bits in Diagrams A and B, it becomes clear how the trimmer would move to the left.

As the trimmer is being pushed forwards by the operator, it will naturally trend left while cutting the groove. However, the fence prevents it from veering off course and keeps the trimmer moving straight.

With the fence on the left side, the trimmer will move in a straight line and creates a perfectly straight groove. The rule is simple: ' Always attach the fence to the left side of the trimmer's direction of travel'. This is the golden rule!

On the other hand, in Diagram B, the fence is on the right side. While the trimmer moves forward, its tendency to curve left makes it constantly pull away from the fence. As a result, you would need to forcibly keep the trimmer pressed against the fence, which makes achieving a straight groove difficult. In other words, attaching the fence on the right is incorrect.

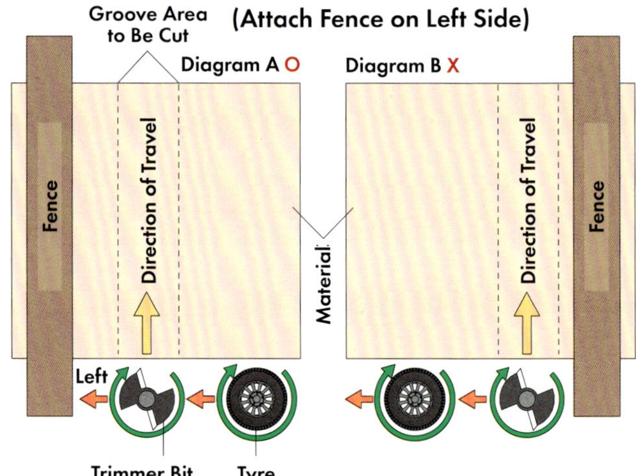

◉ STRAIGHT GUIDE ON THE TRIMMER'S RIGHT

The photo shows a trimmer, with a straight guide attached, performing a partial rabbet cut. Despite the trimmer's tendency to curve left, there is no fence on its left side. Instead, the straight guide serves to keep the trimmer moving straight. This results in a straight cut. If this guide were on the left side of the trimmer, it wouldn't function properly as a guide. The straight guide should be attached to the right side of the trimmer.

Normal Cuts and Climb Cuts

There is a fundamental rule when trimming around the perimeter of any material. It involves the direction in which you move the trimmer, referred to as a normal cut and climb cut.

To make things easier to understand, we will use illustrations of a circular saw. Normally, when you hold a circular saw and cut material by hand, you cut as shown in the ' normal cut' diagram below. The blade rotates clockwise, and moving it in the direction of travel helps cut the material. Pay attention to the relationship between the direction of the blade and the material. The blade hits the material and cuts while 'pulling' the material up into the base plate. This is the correct way to cut, and it is called a 'normal cut'.

On the other hand, when the direction of rotation is the same but the direction of travel is reversed, we have a ' climb cut'. In a climb cut, the cutting edge of the blade is facing backward in relation to the material (see Diagram 1). If you move the saw in this state, the blade doesn't cut into the material. Rather, it rides up onto the material itself (see Diagram 2). If you think of the saw blade as a tyre, you'll immediately notice that it would try to climb up over the material in this configuration – like going over a bump! Even if the operator tries to hold the saw down, they can lose control, a phenomenon known as 'kickback'. This is very dangerous! It can lead to serious injury, so please exercise extreme caution.

◉ CUTTING WITH A CIRCULAR SAW

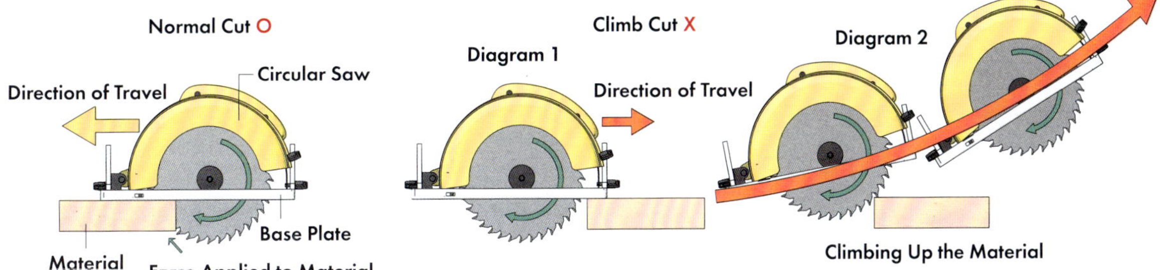

Let's apply the concepts above to trimmer work. The trimmer bit in the lower right diagram is the same as the circular saw on the right, with the cutting edge facing the material. This is the correct direction to move the trimmer, and it's called a 'normal cut'.

In contrast, the trimmer bit above faces backwards, just like the adjacent circular saw. The cutting edge is moving opposite to the direction of travel. This is called a 'climb cut', and it poses the same dangers as with the circular saw. Please do not use this method!

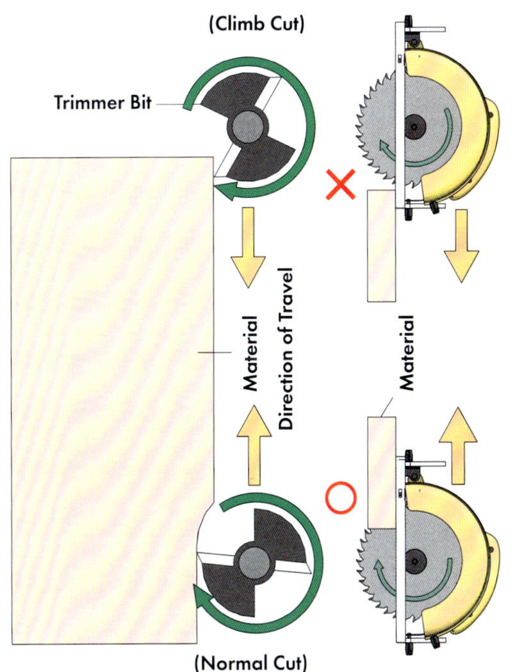

The Right-Hand Rule

When working with a trimmer, it can be difficult to decide on the direction of travel. Here, inspired by 'Fleming's Left-Hand and Right-Hand Rules' (in reference to current and magnetic fields) I've come up with the 'Trimmer Right-Hand Rule' to help clarify things.

1 Butt your right thumb up against the material to be cut.
2 Extend your right index finger. The direction of your index finger is the desired direction to move the trimmer.
3 The same applies when working in a cut-out section. If you butt your right thumb against the area you want to cut, the direction of your index finger indicates the direction of the trimmer.

Now, let me give you a tip. When processing the perimeter of the material, as shown in Figure 2, there is an actual starting order. First, begin by cutting the end grain. Follow the sequence: end grain → edge grain → end grain → edge grain. You will frequently encounter chipping in the final moments of each end grain cut. However, the following edge grain cut will remove those chipped areas. This results in a much cleaner finish.

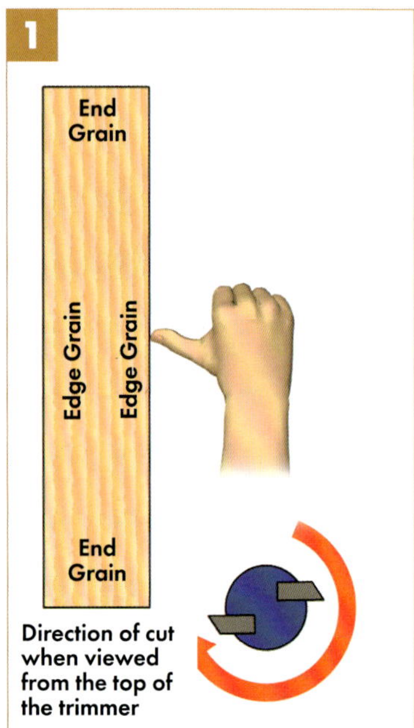

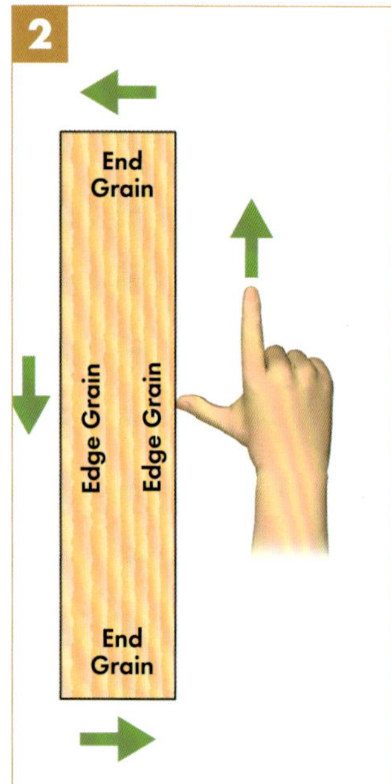

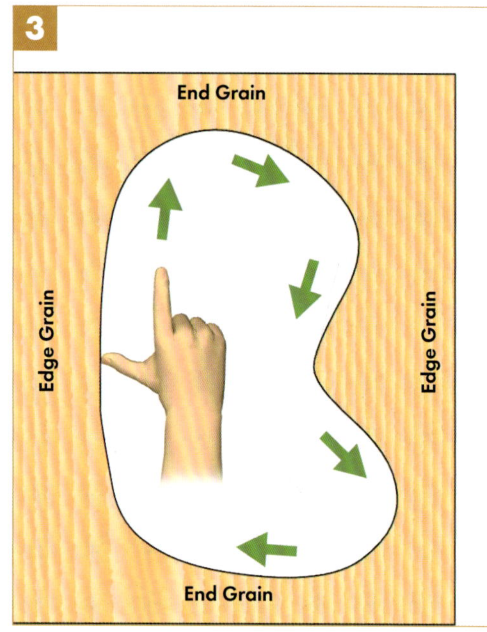

Trimmer Bit Care

When using any tool, sap can build up on the blades. Use a toothbrush or a utility knife to clean off the resin and maintain sharpness. Also, be sure to occasionally clean the shaft of the trimmer bit and the interior of the trimmer's collet chuck.

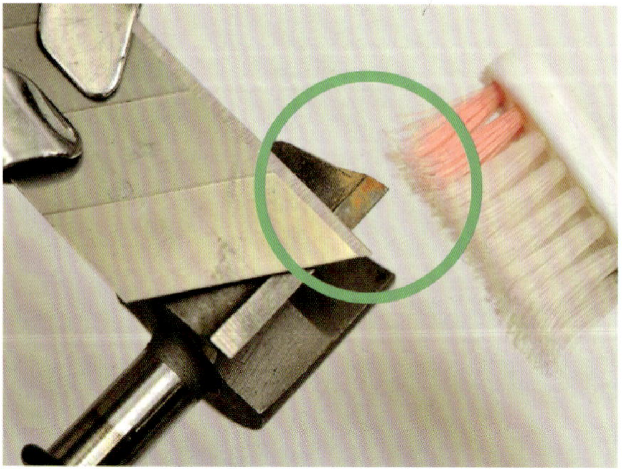

Templates and Template Guides

You can use a trimmer bit with a bearing or a template guide (shown in the photo at right) to create as many duplicates as you like. In my experience, making the template out of ⅜in (10mm) plywood provides a little extra thickness, which makes the template more convenient. However, as with any template guide, there will be a slight distance between the template itself and the trimmer bit when they are aligned. That means you won't be able to create an exact copy of the template. This is referred to as the 'offset' and you need to account for it when making your template. A detailed explanation of the template guide can be found on pages 26–27.

EXAMPLES OF DUPLICATE AND GROOVE CUTTING USING TEMPLATES

CUTTING DUPLICATES

Flush Cut Trimmer Bit

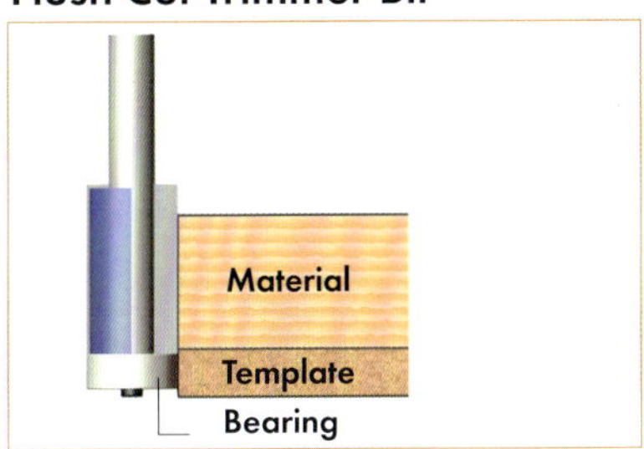

Aligning the bearing at the tip of the flush trim bit with the template allows for easy template duplication.

Pattern Bit

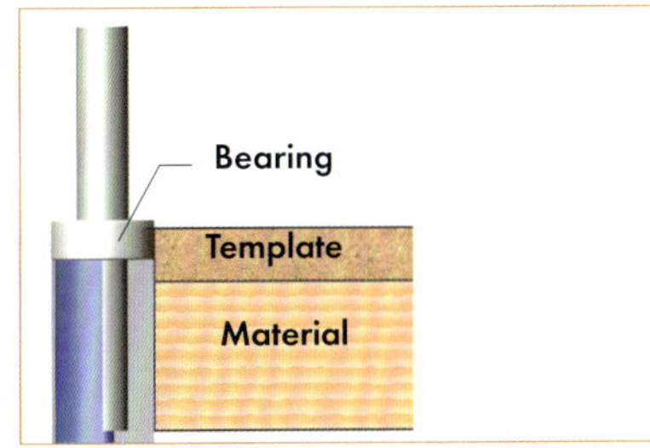

Pattern bits have a shaft located bearing, above the cutter. This means the template is above the target material. Similar to flush trim bits, templates can be easily duplicated.

How to Use a Template Guide

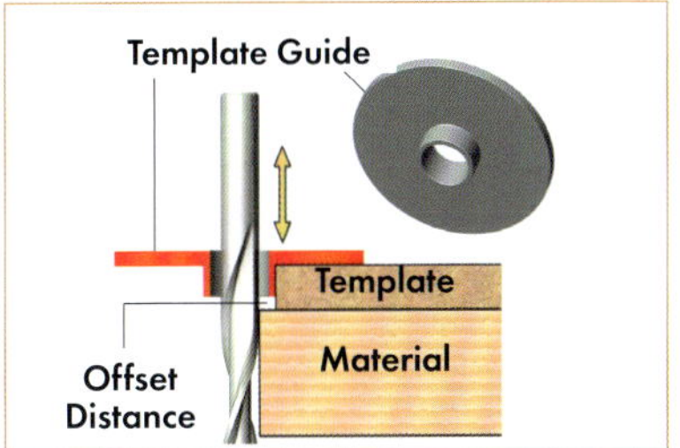

Template guides allow you to use a template even with bits that don't have a bearing. Note, however, that you must always consider the 'offset'.

GROOVE PROCESSING

Using a Template Guide

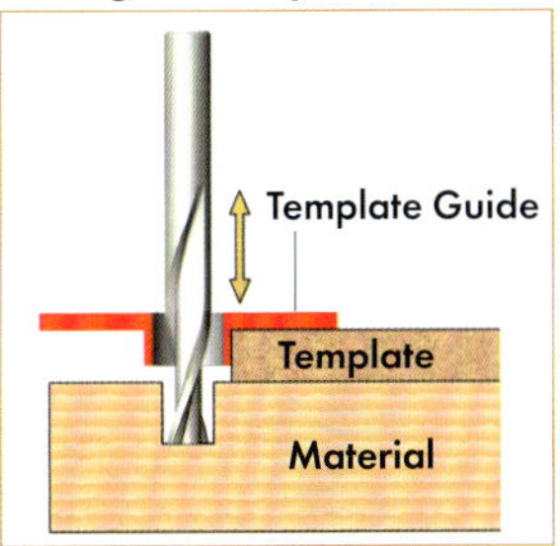

Setting the bit's extension, and aligning the template guide with the template, allows you to create grooves.

Pattern Bit O

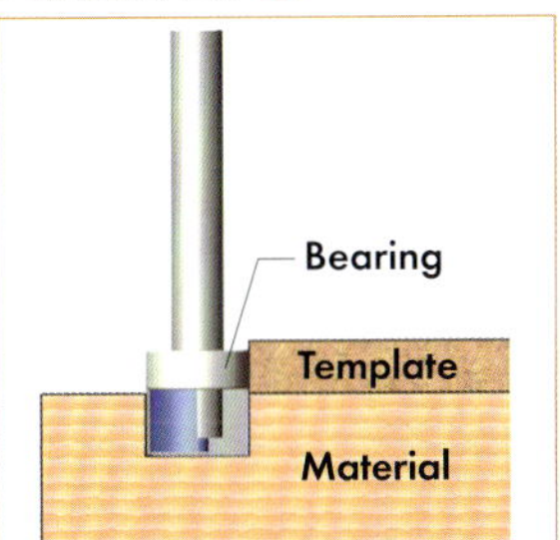

With a short-bladed pattern bit you can process shallow grooves with a thin template.

Pattern Bit X

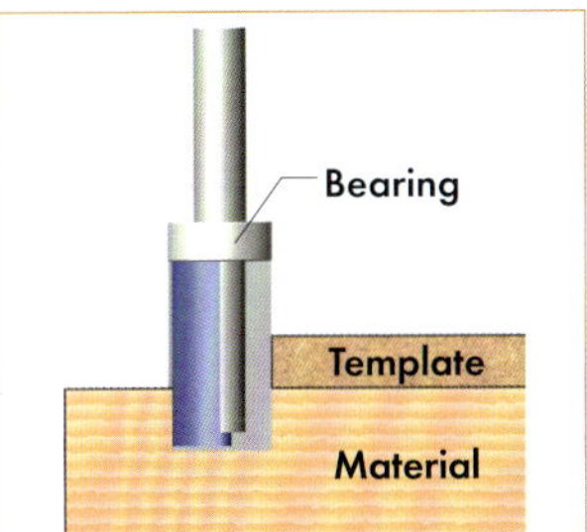

If blade length is greater than the combined groove depth and template thickness, it can't be used for groove processing because the bearing won't contact the template.

Pattern Bit X

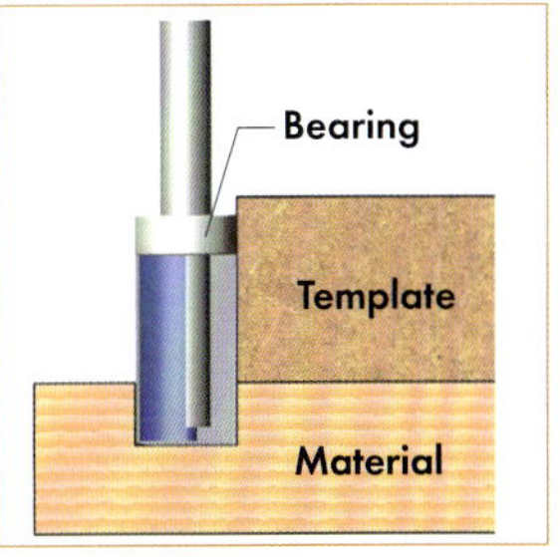

One possible solution to this problem is to make the template thicker, as shown above. But, that's not very practical.

Differences Between Trimmers and Routers

This book is dedicated to trimmers, but I would also like to touch on routers. I often receive questions about the differences between routers and trimmers, so I will take this opportunity to provide a more detailed introduction to routers. Instructions found in this book on the various ways to use a trimmer should help create a clearer understanding of the differences.

Once you are proficient with the trimmer, the next step is the router. The difference between a router and a trimmer isn't just motor size. Routers can use two types of cutters (bits) with different shaft diameters – thin and thick – allowing for a wider variety of bit types. This expands the range of cutting. Routers are quite versatile when compared to other electric tools. They also offer the advantage of stability because you can grip them with both hands.

Currently, plunge routers are quite popular. They have a built-in spring that allows for easy up-and-down movement of the motor. You can set the bit depth of a plunge router using a stopper and can return to the original depth by allowing the motor to rise up. This

allows for operations such as creating long holes or grooves in the middle of a piece of material, while the bit is still rotating! This plunge feature is a definitive difference between routers and trimmers.

In some parts of the world, such as the United States, routers are widespread and trimmers aren't as common. One reason for this is the high skill level of woodworking enthusiasts. As already discussed, routers – with their plunge function – are well-suited for tasks such as mortising and grooving. Since many woodworking enthusiasts perform advanced operations – like mortise joints – using routers, it is rather understandable that routers are more common than trimmers.

A combination-type router, as shown in the photo below, is one that has a single motor and can be attached to either a plunge base or a fixed base. While it may not be superior to a standalone plunge router, it offers the convenience of using the plunge base as a handheld router and the fixed base as a router table.

Template Guide

Another way to expand the range of router applications is to use a 'template guide' (see page 23). This guide is attached to the router's base and is typically used in conjunction with straight bits or spiral bits.

A template guide is a tool used for processing materials with a template. It could also be referred to as a guide bushing (see Diagram 2). Template guides come in various sizes, allowing for differentiation based on intended use.

For processing with the commonly known templates, we use pattern bits or flush trim bits that have bearings attached. These bits have the same diameter for the cutting edge and the bearing, so by aligning the bearing with the template, as shown in Diagram 3, we can easily produce duplicates. However, this method doesn't allow for vertical adjustments of the bit, as the bearing would move away from the template. The solution to this problem is the template guide. As shown in Diagram 2, it acts in place of the bearing while allowing the bit to move up and down. We use straight bits or spiral bits that do not have bearings. However, it's important to understand that there will be an offset distance between the template and the bit. As you can see, a template guide is essential when using a plunge router.

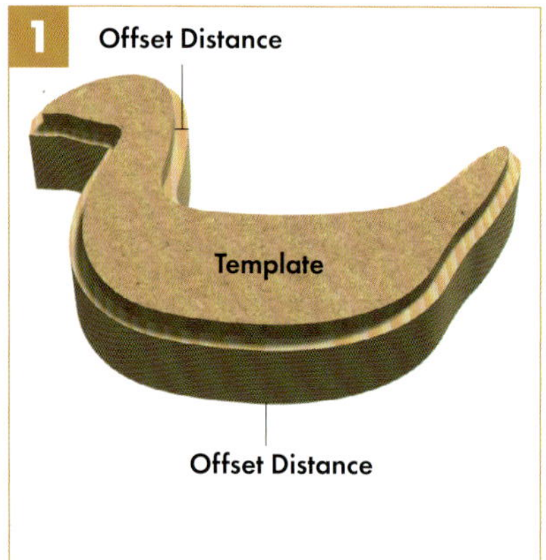

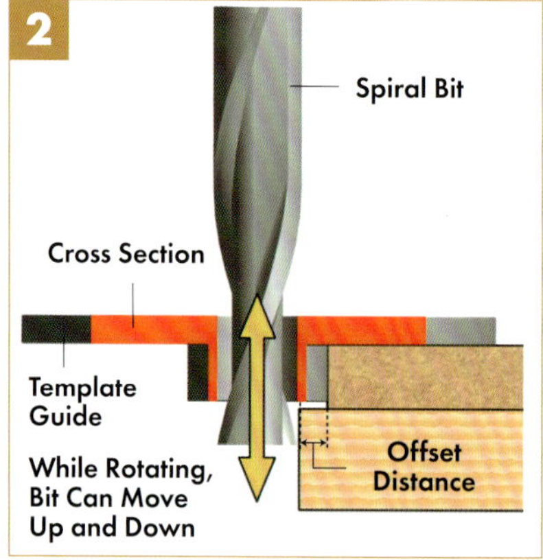

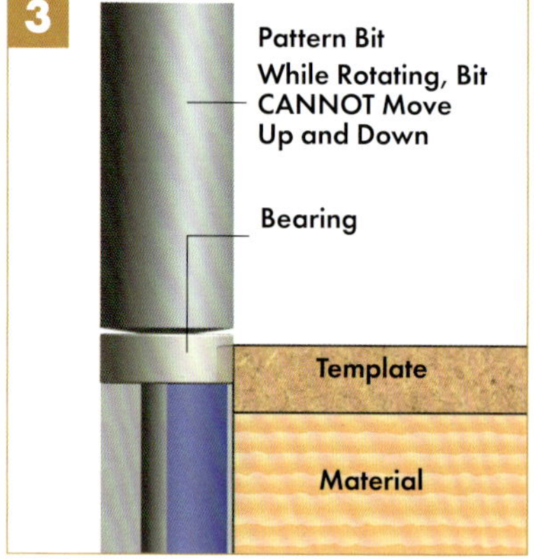

EXAMPLES OF TEMPLATE GUIDE USE

1 One commercially available guide is the dovetail jig. Various dovetail jigs, for use with a router, are sold by different companies. Many of them use a template guide (see opposite for the MIRAI dovetail jig).

2 The second example is a homemade jig. Various jigs can be designed at home, but here we'll introduce a mortise cutting jig. This is a representative example that utilizes a template guide and the plunge function of the router.

3 Template guides have different shaped mounting parts depending on the manufacturer. So, it is essential to purchase the manufacturer's dedicated parts. In addition to these proprietary options, there are also universal template guides, which are so-called 'standard' template guides. There are a variety of sizes available. The OD, or 'outside diameter', is the primary reference point.

Outer diameter dimensions are often labelled 'OD' (Outside Diameter), but this isn't absolutely necessary.

For example, a 3/8in outside diameter can be represented as 3/8". Please refer to the size chart in the lower right for universal sizes in millimetres.

Some manufacturers sell displayed in millimetres. If you pay close attention, you'll notice that some bit sizes actually match up. The sizes marked with an asterisk (*) in the chart below right are compatible.

You can also find universal template guides at some online stores that can be attached to any manufacturer's router. This base has many mounting holes. You can use it by replacing the plastic base that comes with your router.

Universal template guides are available in various sizes. They can often be attached directly, or via an adapter to different manufacturers' routers.

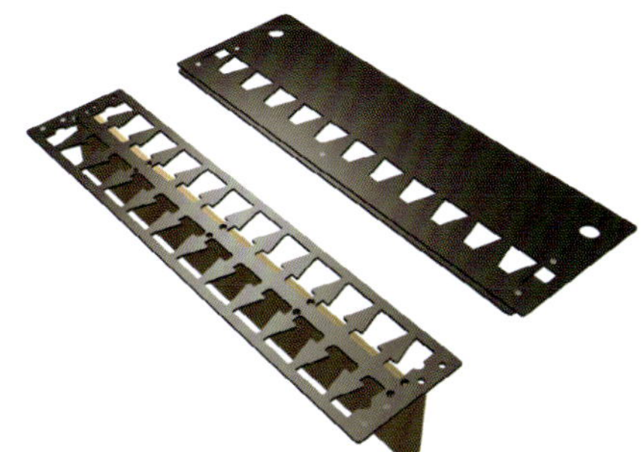

MIRAI Dovetail Jig

Below Left: Through Dovetail Joint.

Below Right: Using a MIRAI Dovetailing Jig. The template guide and rotating bit are visible.

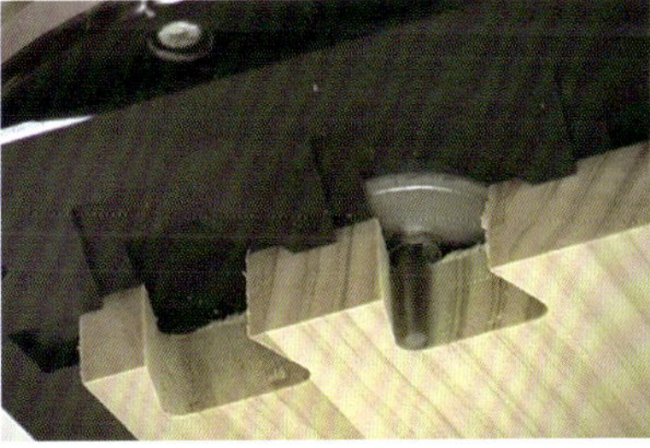

SIZE GUIDE

UNIVERSAL TYPE IN EACH SIZE (OD)

5/16in (7.93mm)

* 3/8in (9.52mm) Hitachi/Makita

* 7/16in (11.11mm) Hitachi

* 1/2in (12.7mm) Hitachi

5/8in (15.87mm)

3/4in (19.05mm)

51/64in (20.24mm)

(1in = 25.4mm)

MORTISE HOLE JIG

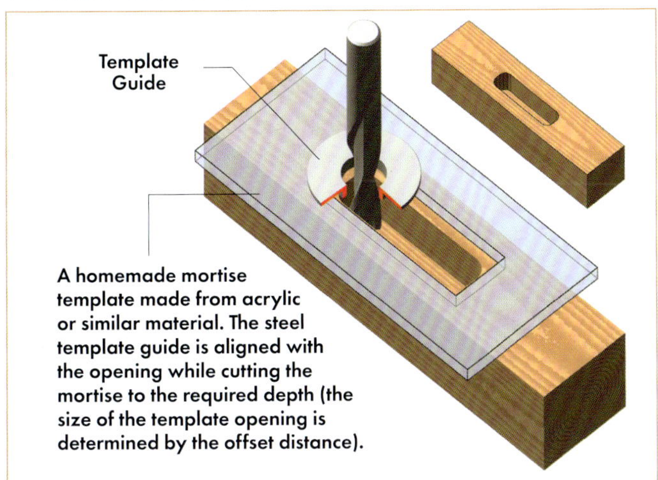

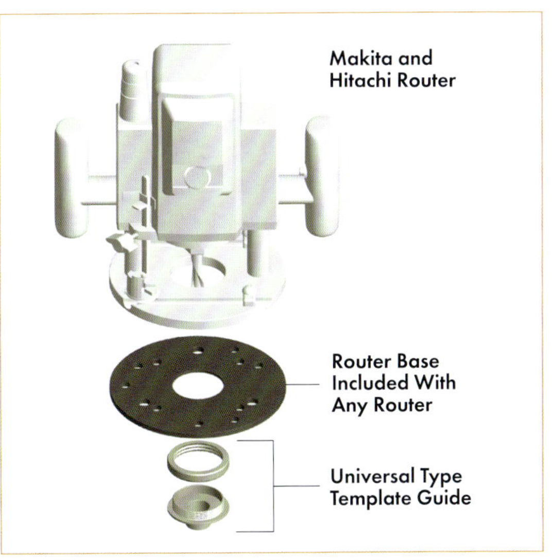

ROUTER BITS

If your router can also accommodate inch-sized bits, variety increases dramatically. Shank diameters are in in inches in the United States and millimetres in other countries, with shank sizes of 12mm or ½in (12.7mm) and 6mm or ¼in (6.35mm). Collet chucks should thus be obtained directly from manufacturers that match these shank diameters. This will allow you to use both millimetre and inch sized bits. The most basic bits are straight with a larger shank diameter. They are versatile and useful for various applications, such as cutting grooves. Photo 1 shows a spiral bit that functions similarly to a straight bit, but provides a smoother cut. The bit in the photo has a twist direction that is the reverse of the usual. It is known as a down-cut bit. This excellent bit reduces the likelihood of fraying on both edges of the groove during processing. The standard twist type is called an up-cut spiral bit. Spiral bits are made entirely from hard tungsten carbide. By contrast, standard bits have carbide cutting edges brazed to a steel body. There are also bits with very large cutting edges in the thicker shank diameter category. It's important to note that some bit manufacturers recommend using them only with a router table. The rail/style bit shown in Photo 5 is designed for processing joint areas on doors and is intended for use with a router table. Other examples include large 45-degree bits (Photo 2), finger joint bits for laminating wood edges (Photo 3), and moulding bits (Photo 4), which are also specifically designed for router table use.

HOW TO CHOOSE A ROUTER: FIRST-TIME PURCHASERS

■ The current mainstream router is the plunge router. If the built-in spring is too strong, it can be difficult to push down, making it difficult to use. Be sure to check the actual product before purchasing.

■ If the top of the router is flat, it will remain stable even in an inverted position, making it easier to change bits and attach or detach template guides.

■ When cutting grooves, it might be necessary to make multiple passes to reduce load on the bit. In such cases, a rotary (turret) depth stopper that is divided into several steps makes adjustments easy. Additionally, it's very important that the depth setting includes a fine adjustment feature. Choose a model that allows for easy, fine adjustments.

■ If there is a speed controller on your unit, you can adjust rotation speeds according to the bit diameter. Larger bits should be used at lower speeds!

■ Having a soft start feature allows for smooth initial cutting.

■ Be sure to purchase a chuck for inch-sized bits as well.

■ Once you become accustomed to using a handheld router, consider converting it into a router table. If you decide to convert your router, removing the built-in spring will make depth adjustment easier.

2

Chapter 2

Handheld Operation

Points on Handheld Operation

'DIY Base Plate' to Improve Machining Accuracy

Handheld operations become much more stable with a large DIY base plate. When processing edge material with a commercially available base plate, more than half of the base plate is suspended in the air. This makes it very unstable. On the other hand, using a DIY base plate reverses this situation and provides far greater stability. When you actually use the DIY base plate, you can clearly feel the difference in stability.

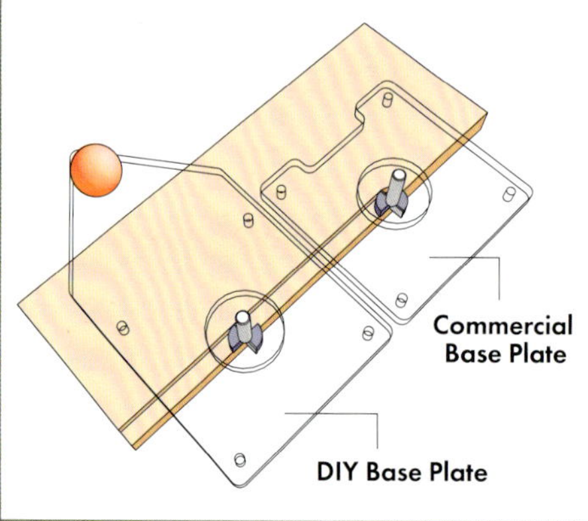

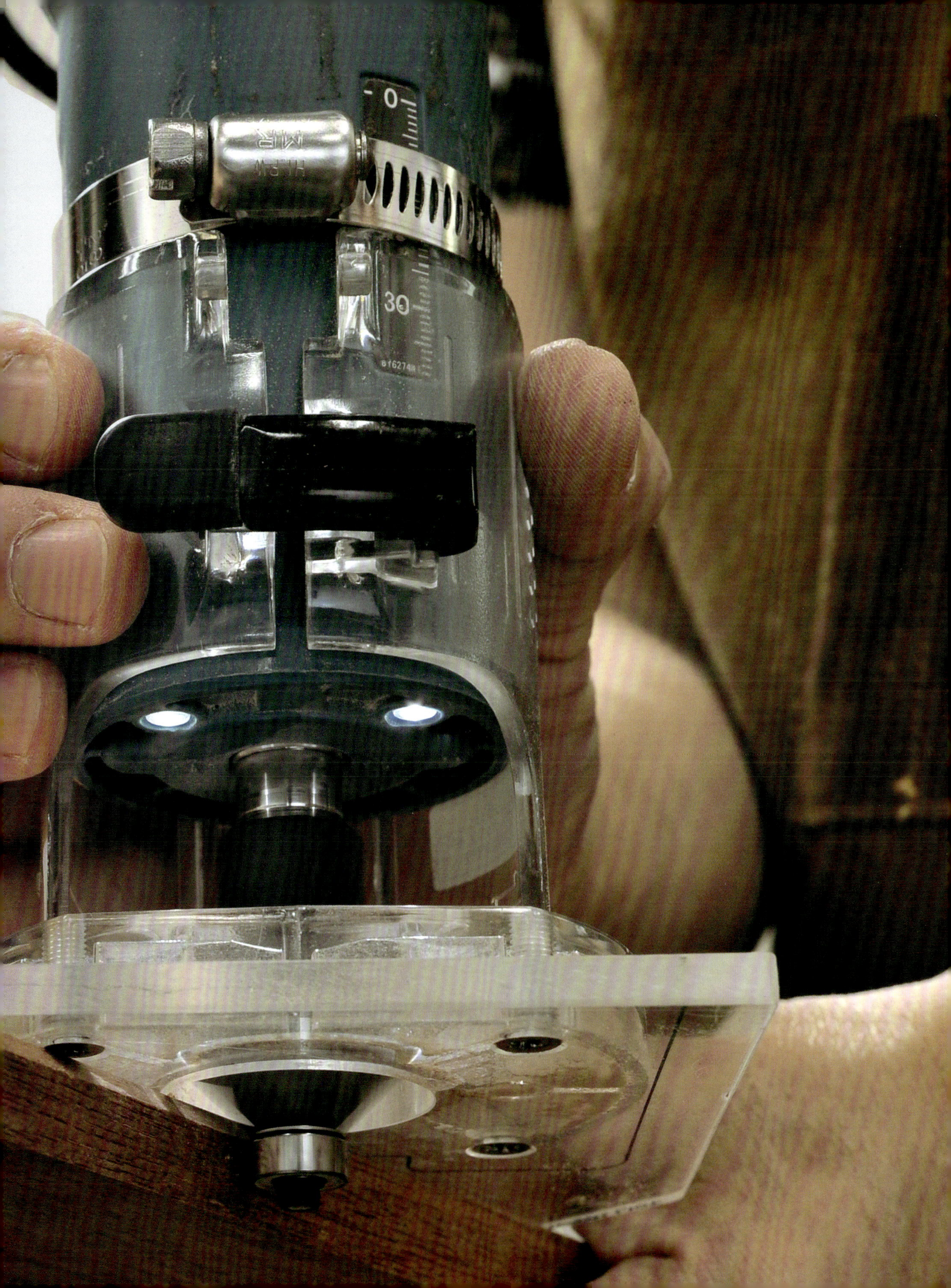
0
30

How to Make a Base Plate

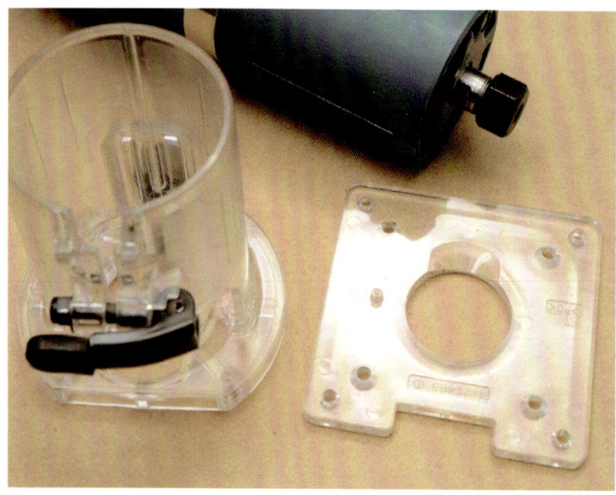

1 Most trimmers have a detachable transparent housing (left) and a base plate (right).

2 Remove the factory-installed base plate and replace it with your large homemade acrylic plate. The acrylic plate is ¼in (6mm) thick. In the photo, the mounting holes are already drilled. These are through holes, so there's no need to tap threads. However, it should be noted that countersinking the holes will be necessary later. Next, I will explain how to position the holes.

3

Secure the factory-installed base plate to the acrylic sheet using double-sided tape. Use the existing holes as a guide and drill holes. Measure the factory base plate hole diameters with callipers and drill the holes to match. In most cases, M4 countersunk screws are used, so the hole size will likely be slightly larger than 5⁄32in (4mm). Also, mark the drill hole positions with a marker.

DRILLING THE BIT HOLE

4

Drill several holes inside the marked circle. Then use a jigsaw to cut between the holes to remove unnecessary sections.

◉ COUNTERSINK BITS

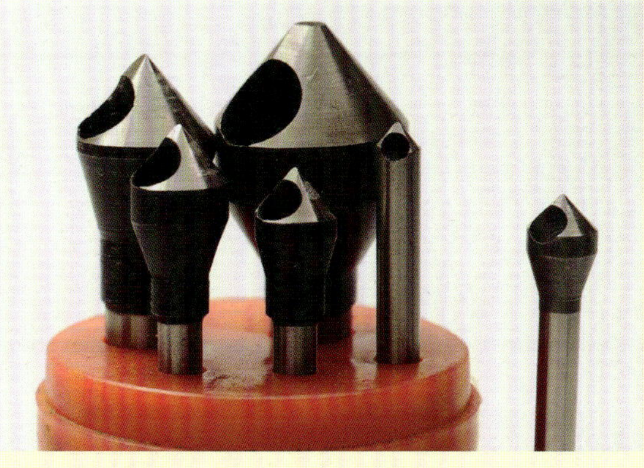

A countersink bit is necessary to allow screw heads to sit flush with the surface.

5 Reattach the factory base plate to the acrylic sheet and use it as a template. Attach the flush trim bit to the trimmer and finish the bit hole. Make sure the bearing part touches the template.

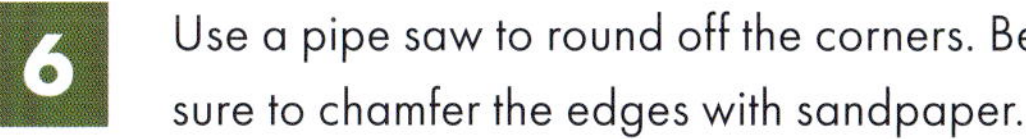

6 Use a pipe saw to round off the corners. Be sure to chamfer the edges with sandpaper.

7 Attach knobs, or grip balls/drawer handles, and it's complete.

The 'Edge Trimmer' for Trimming Edges

When an auxiliary base is attached to the DIY base plate we obtain a new function: Trimming plywood edge banding (see diagram on page 35).

Edge banding that is a few millimetres greater than the thickness of the plywood is added when manufacturing edge-banded plywood. Later, the excess (or offset) can be trimmed off smooth from the edge. Using an edge trimmer at this stage will provide a nice, clean finish.

Attaching an auxiliary base under the DIY base plate creates an offset. Lower the trimmer bit to the point where it barely makes contact with the plywood. If your bit has rounded corners, like a dish cutter for example, there will be less likelihood of leaving cut marks on the material. Make the tip of the auxiliary base triangular, as shown in the structural diagram on page 35. This makes it easier to use.

Edge-band trimming with an edge trimmer.

If we remove the motor, you can see that the edge banding has been trimmed flush with the plywood surface.

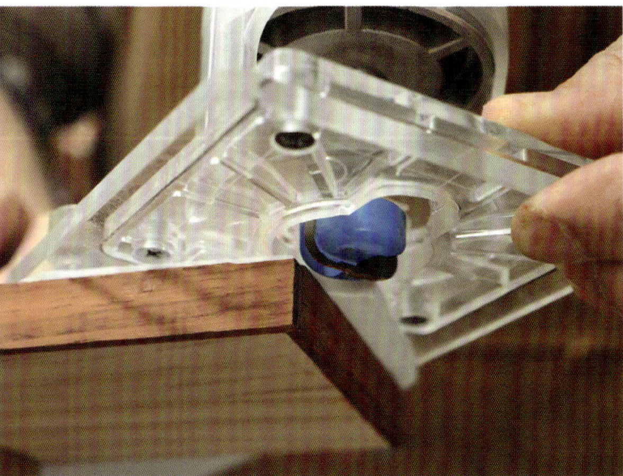

View from below. Be sure to cut with the edge trimmer by continuously making circles, moving from outside the surface into the edge banding and plywood.

STRUCTURAL DIAGRAM

DIY Base Plate

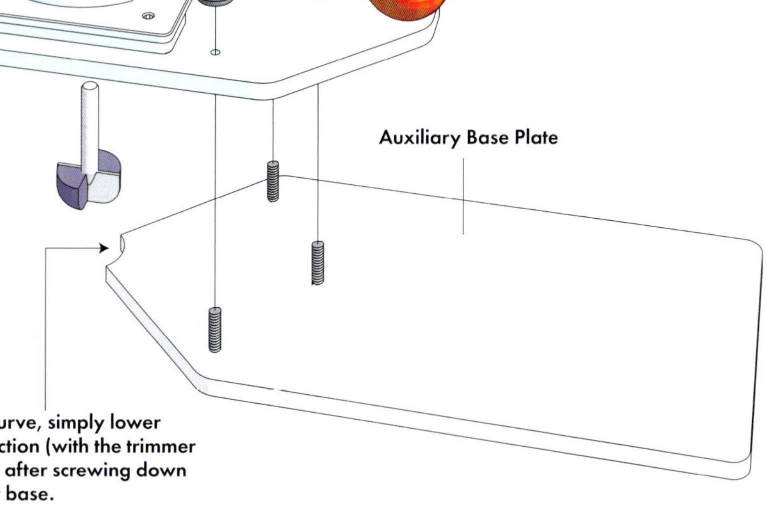

To trim this curve, simply lower the motor section (with the trimmer bit attached) after screwing down the auxiliary base.

OFFSET CORRECTION/MISALIGNMENT

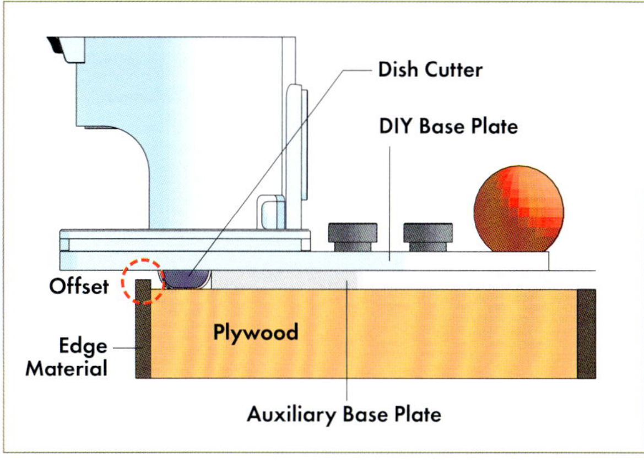

Aligning the dish cutter with the plywood surface allows the protruding edge banding to be trimmed away. This results in a nice, flat surface. The effect is similar to planing.

CUTTING DIRECTION

Edge Material

Plywood

Avoid chipping by bringing the blade in from outside the edge of the material.

ADHERING EDGE BANDING TO PLYWOOD

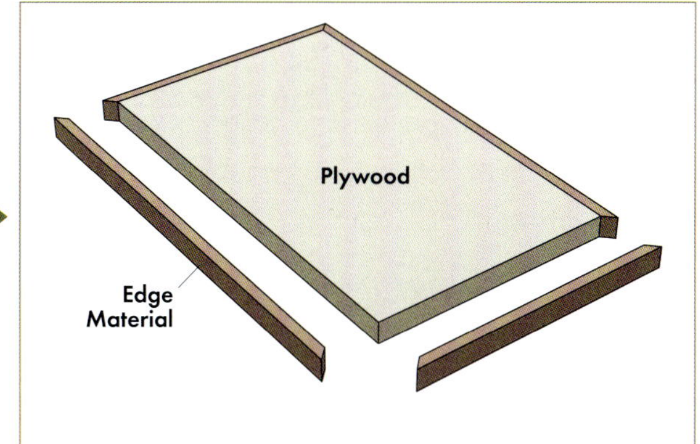

'Trimmer Planer' for Correcting Warping and Determining Thickness

This jig is used to correct wood warping, twisting, and other deformations, and to finish to a required thickness. As shown in the diagram below, a trimmer fixed to a U-shaped carriage is moved back and forth on two rails – and side to side – to flatten the wood.

The photo on the right shows a workbench surface in use. The two visible 1×2in (25×50mm) rails, and the material to be planed, are secured with double-sided tape. If the material has warping and wobbles when placed on the workbench, use thick double-sided tape or wedges to stabilize it before planing.

One important point to note during construction is that the bottom of the carriage must be strong enough to prevent sagging. During planing, the carriage will be pressed down with a certain amount of force, so it must be strong. If the risers are too narrow, the entire carriage will be more prone to sagging. Please refer to dimensions in the diagram below for correct sizing.

Use a large-diameter bit to flatten wide surfaces.

EAGLE AMERICA(USA) Hinge Mortising Router Bit/Blade Diameter: 1¼in (32mm)/Blade Length: ½in (12mm) WHITESIDE (USA) Threaded Mortise Cutter/Blade Diameter: 1¼in (32mm)/Blade Length: ½in (12mm)

Flattening one side of a piece of western red cedar. Flip the material over and plane the opposite side as well. Once both sides are flat, finish to the required thickness. Make sure that both hands press down on the carriage directly over the rails.

◉ TRIMMER PLANER

Carriage Cross-Section

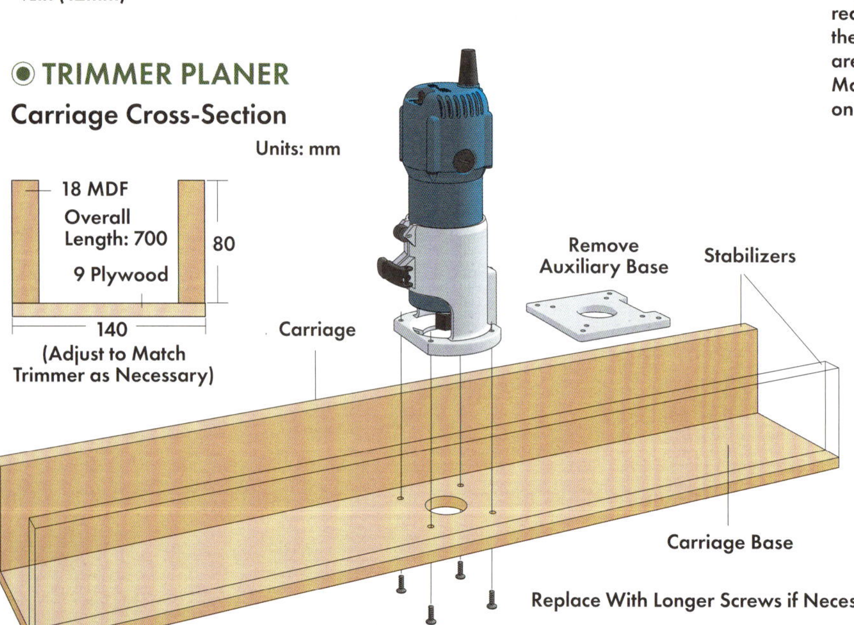

DEPTH SETTING WITH A HOSE CLAMP

You can use the hose clamp as a depth stop. The method here is to replace the thickness of the material you want to finish with the thickness of the acrylic board. Then, use that as a spacer.

Lower the trimmer bit onto the spacer and tighten the hose clamp at that position. Ensure that the hose clamp is in contact with the top edge of the housing (the transparent part of the router). That way, the trim router bit won't slide down while planing. You can now finish the material to the required thickness.

When cutting large amounts of material, it's best to reduce the load on the trimmer bit. To do this, keep the stopper position as it is but reduce the amount of the cut by setting the housing slightly lower; separate the housing and stopper a little bit and then firmly tighten the black housing lever. You should normally cut about 3/64–5/64in (1–2mm), depending on the type of material.

You can see the transparent acrylic spacer under the router. The hose clamp is set so that it contacts the edge of the router housing. Tighten the lever housing to secure the motor in place.

METHOD FOR ANGLE CARVING MATERIAL

Changing the height of the left or right rail means that you can cut material at an angle. This is super convenient when you need to create material that tapers (slopes).

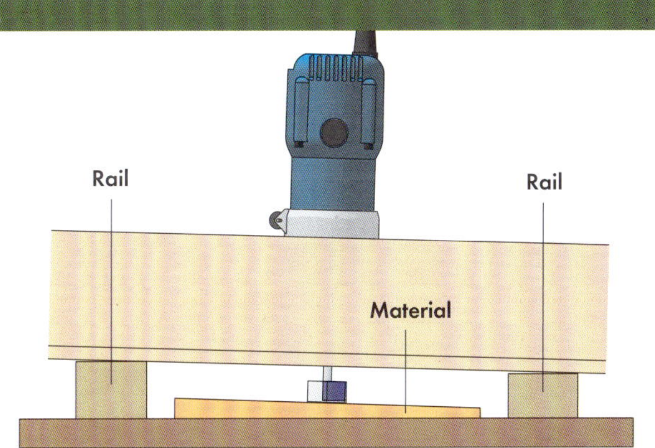

XY-AXIS SLIDING PLANER

The purpose of this jig is the same as above, but it has a slightly different structure. The trimmer isn't fixed and can move left or right on the carriage. This allows for a shorter carriage length. To cut the material, the trimmer moves left-right along the X-axis, while the carriage moves forwards and backwards along the Y-axis. Stoppers are necessary at both ends in order to prevent the carriage from derailing.

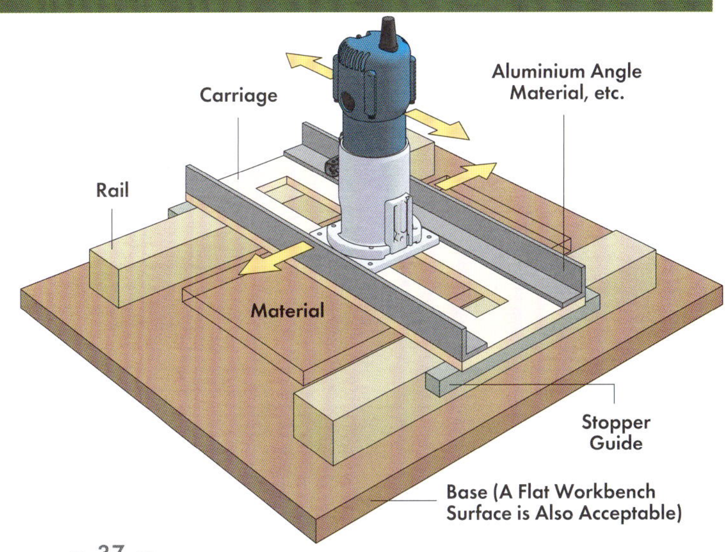

An 'Offset Gauge' for Convenient Fence Operations

Grooves cut with trimmers are often made using a fence. In this process, the distance between the fence and the bit is called the 'offset'. Having a gauge that automatically fixes the offset allows you to quickly and easily set up your trimmer, and to maintain a consistent distance between the pencil line marking the groove and the fence. It's even more convenient if you make this type of gauge for each trimmer bit used. Here, we will make one for a ¼in (6.35mm) spiral bit.

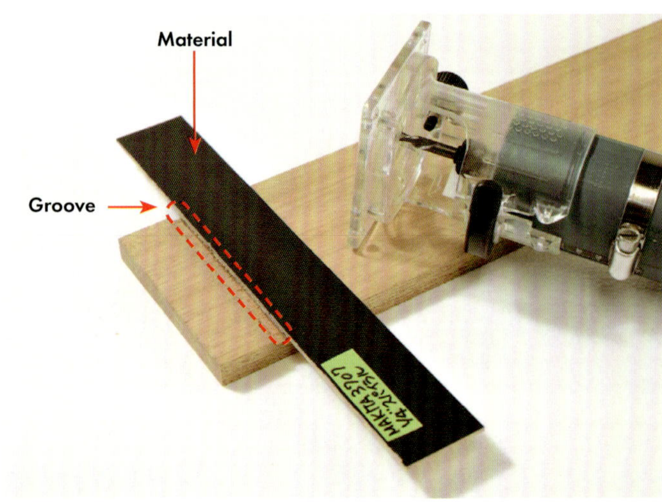

CREATING AN OFFSET GAUGE

Prepare a piece of material (black plywood in this case) that is approximately ¼in (6mm) wider than the distance from the trimmer base plate to the bit. Fix your fence to a workbench and secure the black plywood with double-sided tape. Make sure that the material is snug against the fence. If the material is too thin, it's really prone to warping, so we recommend using 5⁄32–¼in (4–6mm) thick plywood or MDF.

Lower the bit tip below the depth of the gauge material and push the trimmer along the fence. The material that remains will become your offset gauge. Its width will match exactly the distance from the trimmer base plate to the bit. Make sure to keep the trimmer in the same orientation (i.e. 'lever always against fence') each time you use it, otherwise, the offset distance might change.

METHOD

1 Draw a line where you want to cut the groove. Tape the offset gauge down along that line.

2 Butt your fence up against the offset gauge and secure with a clamp. Then, remove the gauge.

3 When you use the trimmer to cut along the fence, a precise groove is produced along the line.

'T-Square Slide Fence' for Convenient Grooving

This is a fence for cutting cross-grain grooves. It's a T-square, so you can easily get the fence set at a right angle by butting it up to your material. There are three different uses for this fence, which makes it quite convenient. As you can see, there's a sliding fence next to the fixed fence. Moving that sliding fence allows you to adjust the groove width. In addition to the main fence, you will also make spacers of various thicknesses from plastic sheets.

You can cut grooves to a maximum width of 12in (300mm).

T-SQUARE SLIDING FENCE DIMENSIONAL DIAGRAM

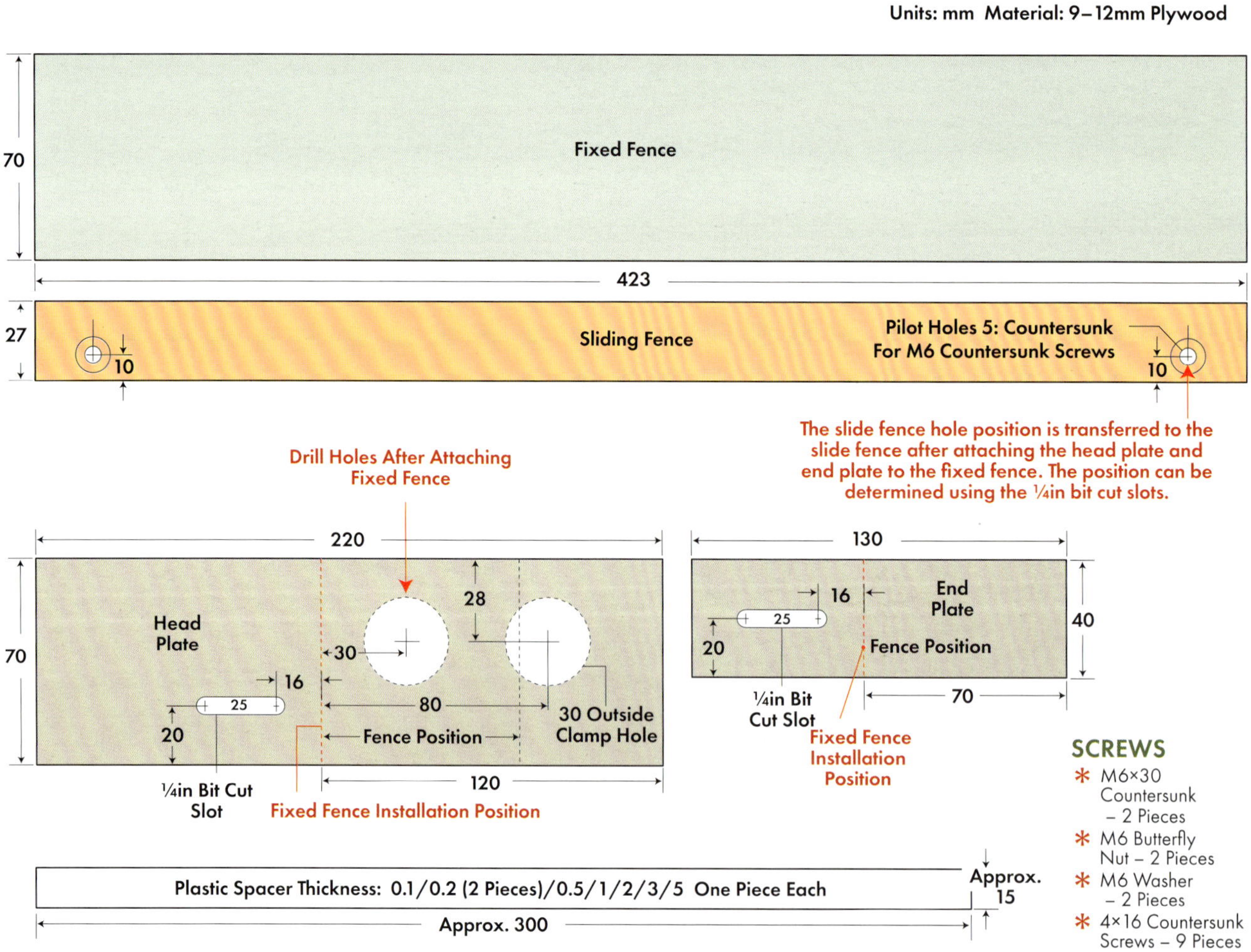

Units: mm Material: 9–12mm Plywood

Fixed Fence
70
423

Sliding Fence
27
10
Pilot Holes 5: Countersunk For M6 Countersunk Screws
10

The slide fence hole position is transferred to the slide fence after attaching the head plate and end plate to the fixed fence. The position can be determined using the ¼in bit cut slots.

Drill Holes After Attaching Fixed Fence

220
Head Plate
70
28
30
16
25
20
80
30 Outside Clamp Hole
Fence Position
120
¼in Bit Cut Slot
Fixed Fence Installation Position

130
End Plate
16
25
20
40
Fence Position
70
¼in Bit Cut Slot
Fixed Fence Installation Position

Plastic Spacer Thickness: 0.1/0.2 (2 Pieces)/0.5/1/2/3/5 One Piece Each
Approx. 15
Approx. 300

SCREWS

* M6×30 Countersunk – 2 Pieces
* M6 Butterfly Nut – 2 Pieces
* M6 Washer – 2 Pieces
* 4×16 Countersunk Screws – 9 Pieces

ASSEMBLY DIAGRAM FOR T-SQUARE SLIDE FENCE

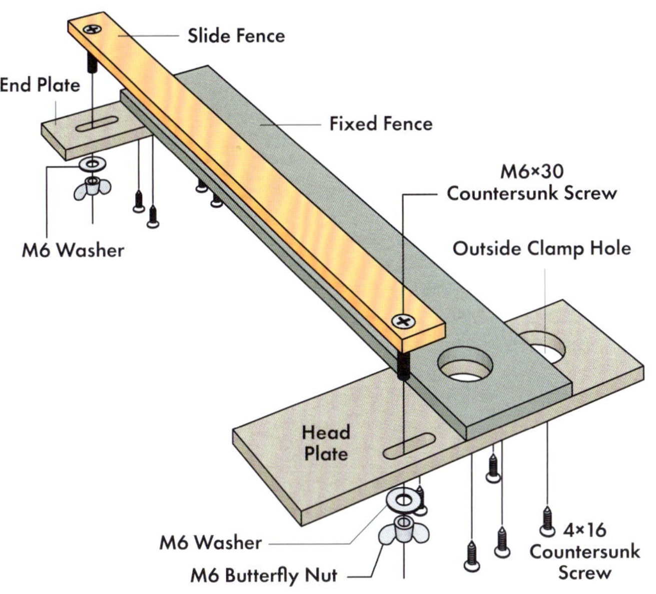

◉ HOW TO SET THE GROOVE DEPTH

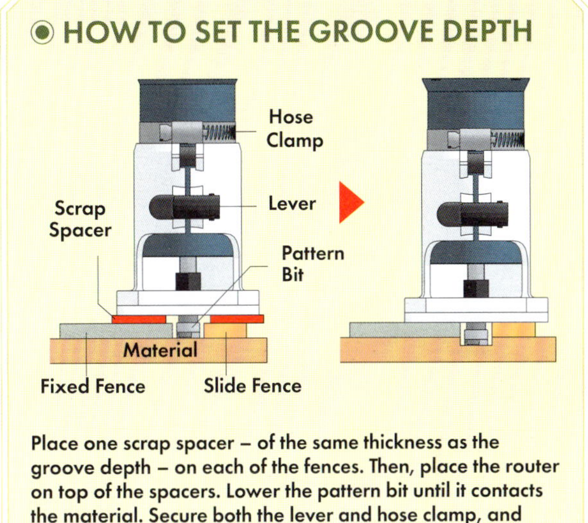

Place one scrap spacer – of the same thickness as the groove depth – on each of the fences. Then, place the router on top of the spacers. Lower the pattern bit until it contacts the material. Secure both the lever and hose clamp, and then remove the scrap spacers to complete the depth setting.

INSTRUCTIONS

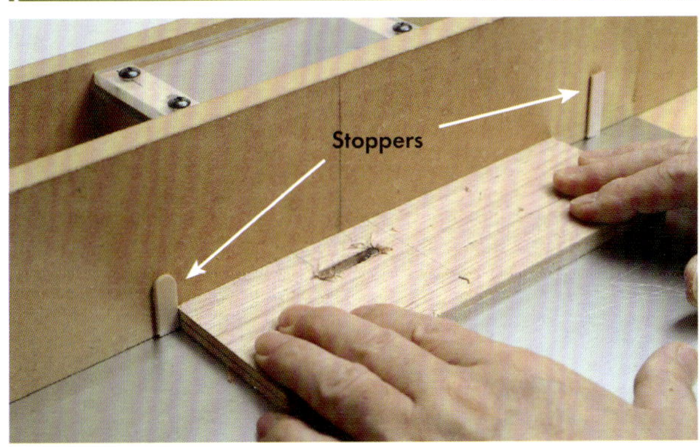

1

Cut materials according to the dimensional drawing. Use a trimmer table to easily cut elongated holes in the end plate and head plate (see page 66). Don't change fence position, just cut both materials. However, the stopper position must be changed. Cut craft sticks to an appropriate length and attach to fence with double-sided tape as stoppers. A ¼in spiral bit works best here. Since ¼in is 6.35mm, it's just the right size for an M6×30 countersunk screw.

2

These photos show how to use a combination square to fix the plate and fence at right angles. Temporarily secure the head plate and fixed fence with instant glue. Then, firmly fix them in place with wood screws. It's best to pre-drill one clamp hole on the head plate beforehand. Attach the end plate to the fixed fence in a similar fashion.

USAGE

1 WHEN PROCESSING GROOVES WITH SAME WIDTH AS THE BIT

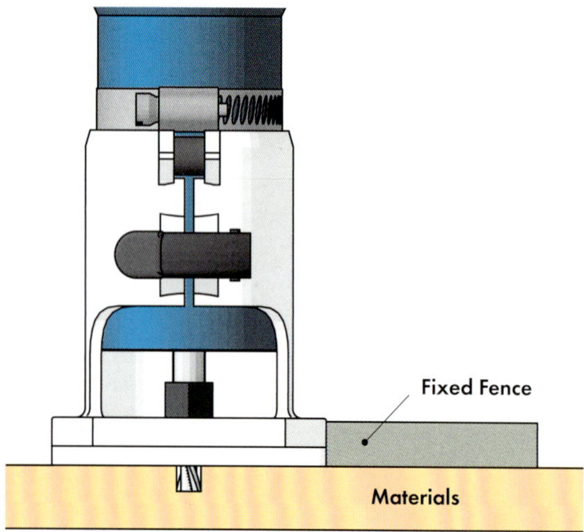

This is the simplest method. Just remove the sliding fence and cut along the fixed fence. The groove width will be the same as the bit's cutting diameter and cannot be changed. The trimmer must be placed directly on the material, so there are holes in the fixed fence that ensure the clamps don't get in the way.

2 USING SPACERS TO WIDEN THE GROOVE WIDTH

Insert a spacer between the fixed fence and the slide fence. Perform the first groove cut. The spacer's thickness should be obtained by subtracting the trimmer bit's diameter from the final groove width. Finally, remove the spacer, close the gap between the fences, and perform the second cut.

Here is a precise dado groove. Down-cut spiral bits are effective for groove cutting because they minimize splintering on both sides of the groove.

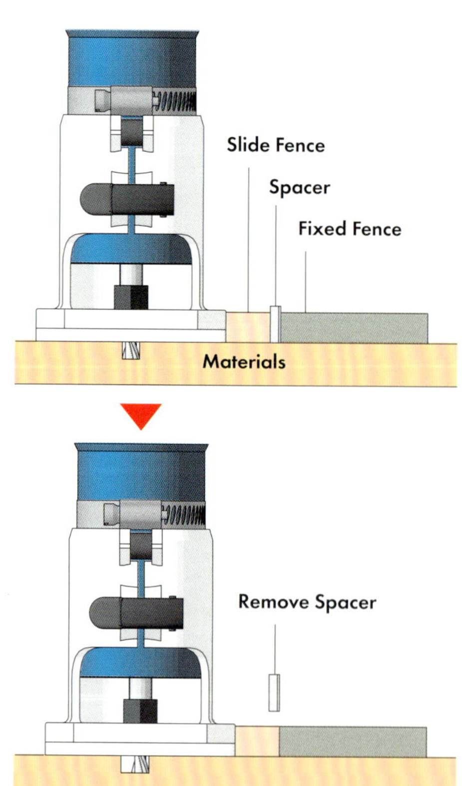

3 CUTTING A DADO GROOVE WITH A PATTERN BIT (DADO JOINT)

Insert the target material between fences to fix the groove width. Use a pattern bit to cut between the two fences. Note that the router is placed on top of the fences during use, not directly on the material. The bit used has both a bearing diameter and cutting edge diameter of ½in (12.7mm), and the cutting depth is very short at ⅛in (3.175mm). Note that this bit cannot be used if the groove width is less than ½in.

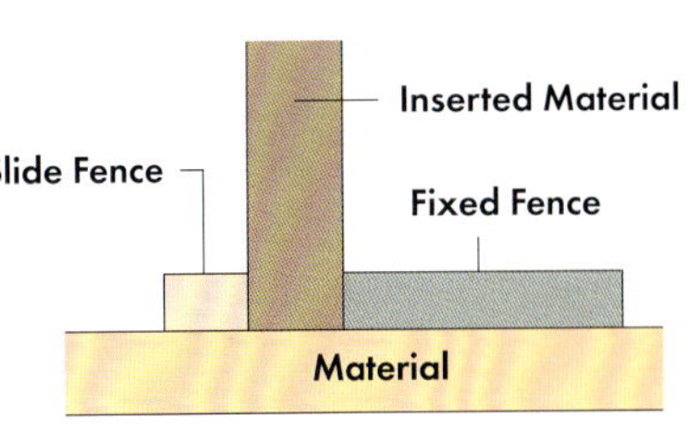

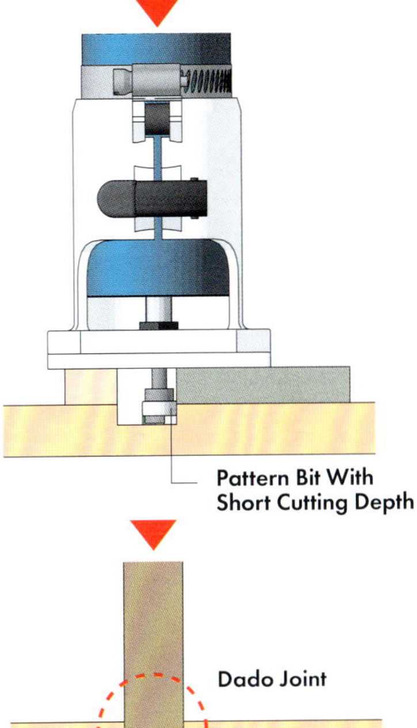

Short cutting depth bit (with a bearing) is most suitable.

Dado Joint

Clamp material that will be inserted into the groove between fixed and sliding fences. Tip: Insert a folded sheet of copy paper to create a slight clearance. This will make the groove width just right. Once the sliding fence is secured, remove the material. Use outer holes for clamping this time. This will ensure the clamps don't interfere with the router. As shown on the left, pencil lines mark out the router's path. Be sure to position the clamps so they don't cross inside those lines.

Use the pattern bit to cut between the sliding and fixed fence. Avoid climb cutting by placing the router so the cut direction forces the router into the fence. Make multiple passes to the desired depth to complete the dado groove.

'Router-style Base Plate' for Working With Both Hands

Wide base plates allow for wider grooves. Cutting rabbet joints becomes easier because now you can cut wide grooves across the grain. Additionally, with handles on each side, precision increases and allows you to perform tasks in a similar manner to a full-sized router. The construction method is the same as the DIY base plate introduced on page 32. Give it a try!

The pattern bit used here is slightly larger in diameter because we are removing more material. It still has a short cutting length and a bearing. As you roughly cut out any unnecessary wood with a chisel, the left and right walls of the groove begin to appear. Align the bearing part of the pattern bit with these walls and you can cleanly flatten the bottom without cutting the sides. We use a short-length bit for precisely these reasons.

There is a ton of material to remove here, so it's best to make several cuts with a saw and roughly cut out any excess material with a chisel. This will extend your trimmer bit's lifespan. As seen in the photo, a craft stick is attached to the saw blade with double-sided tape at the depth corresponding to the groove. This will help maintain a consistent cut depth.

'Edge Guide Jig' for Spline Joints

This jig is attached to the edge guide (aka straight guide) that comes with your trimmer. The second photo on the right shows two pieces joined at a 45-degree angle. There is a groove for the spline cut into the joint surface as well. This spline will increase joint strength. We will use the mitre jig introduced on page 52 or the 45-degree inclined board on page 176 to create the spline joint.

Traditionally, it has been quite challenging to match the groove width with the spline joint thickness. Our edge guide jig solves this problem. By clamping a scrap piece of spline joint material in the jig and performing the first groove cut, and then removing the scrap and making a second cut, the groove width will match the spline joint thickness almost perfectly.

This jig is designed to use a ⅛in (3.175mm) spiral bit. The body features two grooves (pockets) that are ⅛in wide next to the elastic strings. By inserting scrap pieces of the spline into these pockets, the jig will automatically set the groove width. This makes it a highly efficient tool.

Jig use will be explained in the following three pages, while construction details will be provided from page 49. Please note that the trimmer table introduced on page 66 of this book will be used for a portion of the jig fabrication.

METHOD (STEPS FOR MAKING GROOVES)

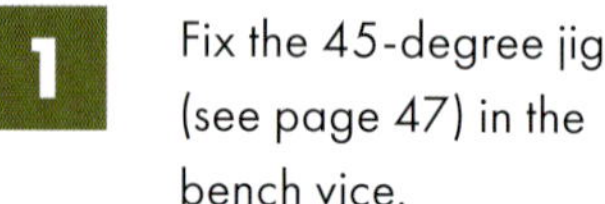

1 Fix the 45-degree jig (see page 47) in the bench vice.

2 Attach the worktable (see page 47). Move the trimmer back and forth to cut a groove.

3 Clamp the material in place. Material end has been processed at a 45-degree angle (see page 54 and following pages).

4 First, insert leftover spline material into the edge guide jig pockets. Setting the trimmer at this position means you are ready to make the first cut.

5 Make the first groove cut; its width will be the same diameter as the spiral bit. Next, remove leftover spline material from the pockets. This will shift the trimmer position by the necessary amount. You are now ready for the second cut.

◉ WORKTABLE AND 45-DEGREE JIG

Please refer to the diagram below to construct the worktable and 45-degree jig. A hand plane and planing board were used to create the 45-degree face, but there are various other methods. Some of these are introduced in this book (see pages 176–179). Alternatively, you can use a saw and a commercial guide to make the cuts.

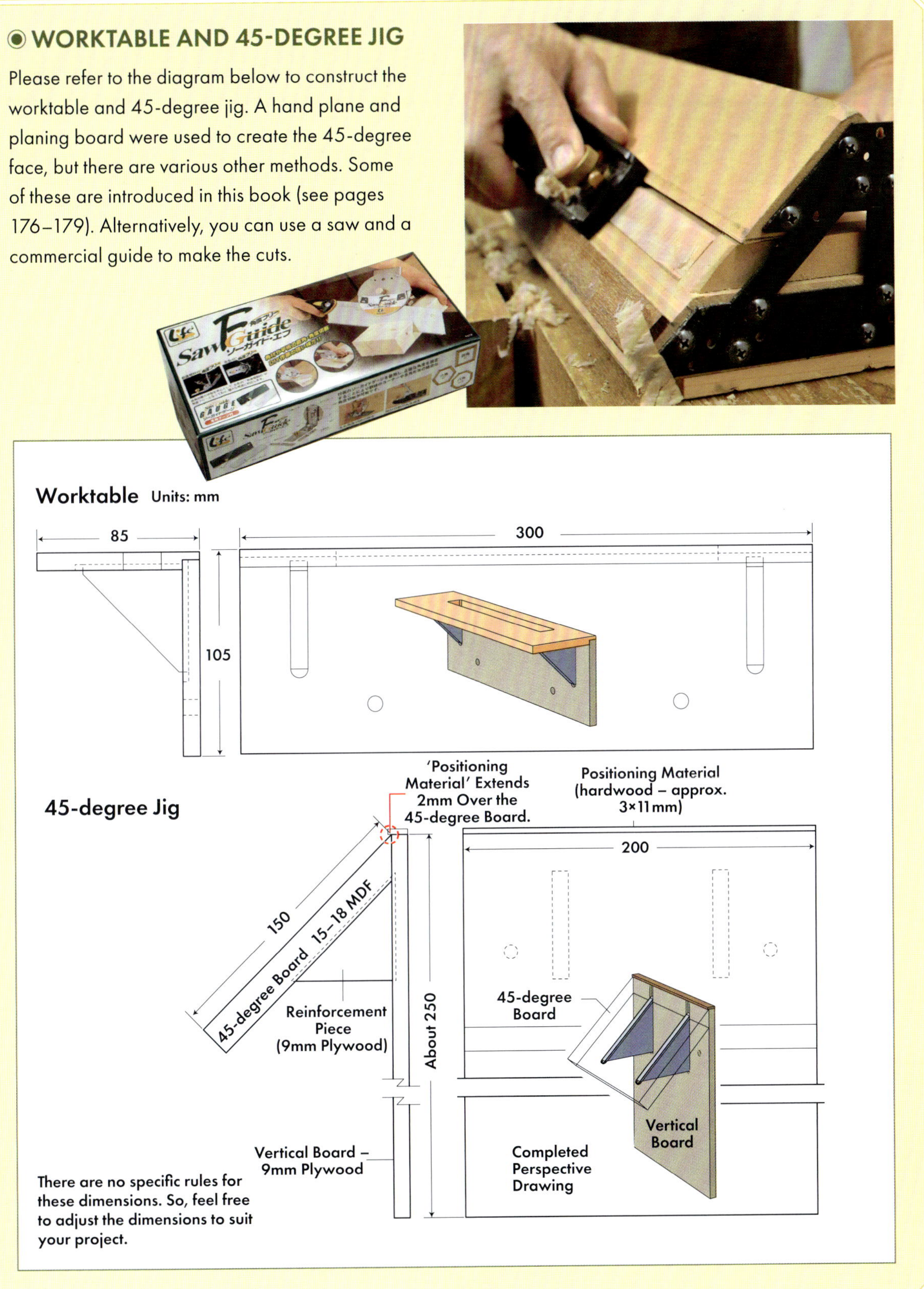

6

The second groove cut is complete. The groove width has now been expanded and matches the spline joint thickness.

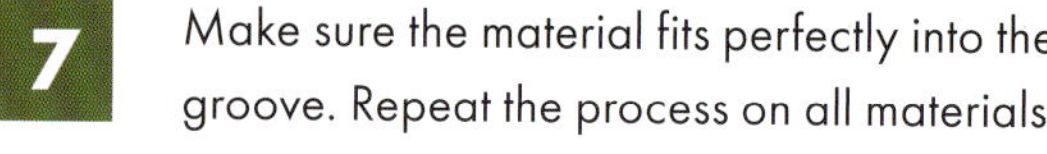

7

Make sure the material fits perfectly into the groove. Repeat the process on all materials.

8

Put two pieces together and measure the spline joint groove width. The photo shows the spline joint inserted into the groove. This completes the process. Even though the groove is visible, you can choose to hide it by not machining the groove all the way to the edge of the 45-degree surface.

HOW TO MAKE AN EDGE GUIDE

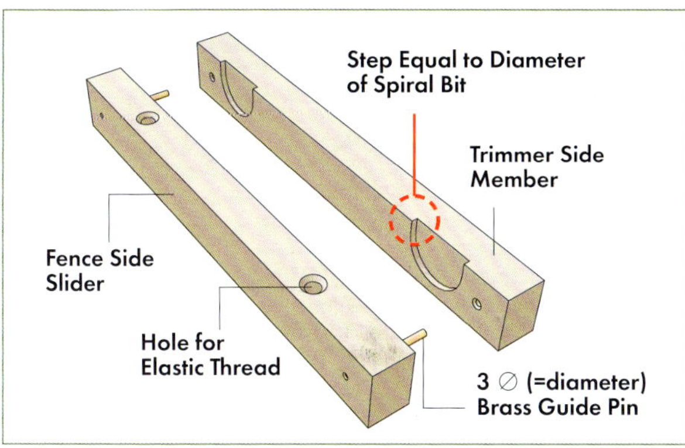

1 The basic edge guide jig structure is quite simple. It consists of two parallel members, secured by two guide pins. Rubber thread is wrapped around the members to prevent separation. This assembly is then screwed onto the trimmer's edge guide.

2 Each part is fabricated as shown in the photo. Brass guide pins have threads cut into them, but you can also just slightly narrow the fence side slider hole and forcefully press guide pins in. If you choose the latter method, you should also slightly enlarge the trimmer side member holes.

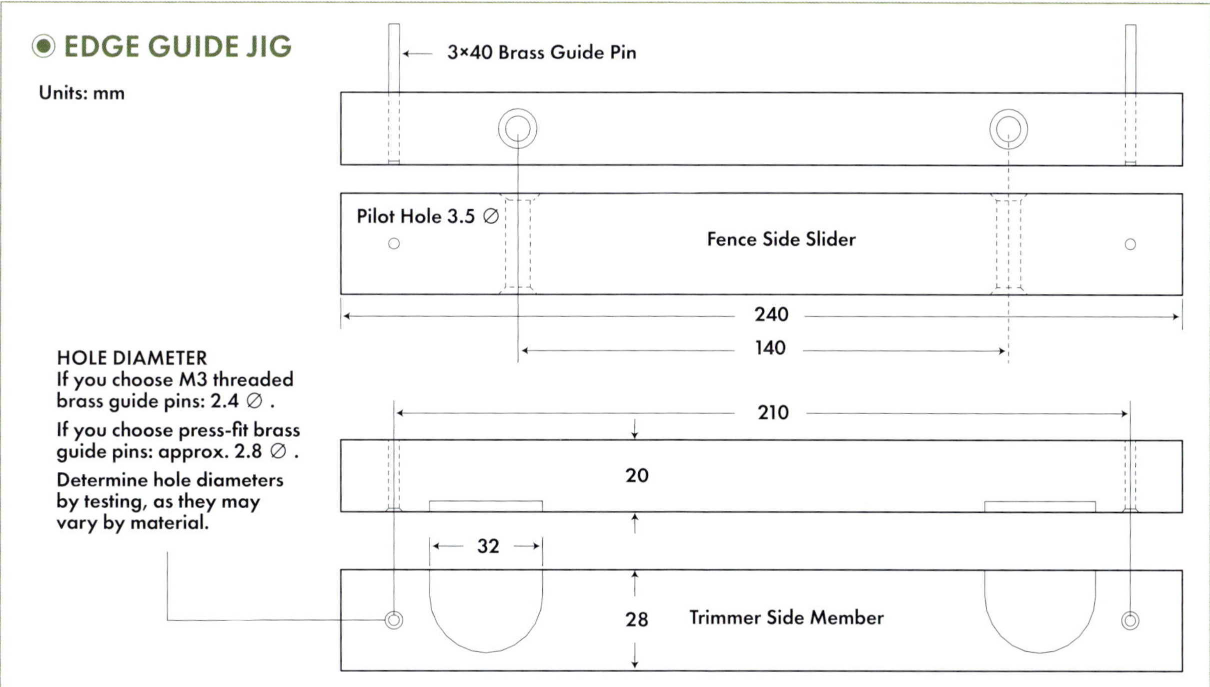

3

Cut a pocket on a trimmer table (see page 66). Set the bit height to spiral bit diameter of 3.2mm. Here, a straight bit with a diameter of 1¼in (approximately 32mm) is used. This should not be done freehand, so a simple jig will be made.

4

This is a simple T-shaped jig. It uses two C-clamps, so gluing isn't necessary. A positioning stopper is fixed with double-sided tape at the tip. The stopper can be moved to match the position where the material needs to be cut.

5

Determine the stopper position. The entire jig moves back and forth, so we need to limit that movement as well. Here, a craft stick is attached to the top plate as a stopper. After cutting one pocket, change the stopper position and process the remaining pocket.

To create the brass guide pin holes, temporarily secure the two materials with double-sided tape (or something similar) and drill. This prevents hole misalignment. First, drill the smaller hole. Then, remove the double-sided tape and enlarge one hole on the trimmer side member using a slightly larger drill bit.

7

Attach the edge guide with screws. Double up the rubber thread and tie it in a knot.

8

The edge guide jig is complete. Its usage for large mitre joints will be explained in the 'Wall-Mounted Box' section on page 206.

Chapter 2 Handheld Operation

Easy Large Mitre Processing

'Large Mitre Jig'

This jig is used for cutting large mitres by hand with a router or trimmer. You can actually cut four pieces at once by clamping multiple pieces to both sides of the 45-degree inclined plates. High levels of precision are required for matching up and gluing each component, so I devised a method – using 'positioning pins' – that prevents any misalignment.

This jig has components – like the 45-degree inclined plates and the top board – that are sandwiched between two side panels. Be careful, if there is any misalignment during the gluing process you won't have an accurate jig. That being said, the components here are glued in place against M4 screw 'positioning pins'. This ensures that the components are securely fixed in the correct position. Epoxy is the adhesive of choice. It comes in a tube that dispenses only the necessary amount. The base resin and hardener are mixed together in the dispenser. Unlike water-based adhesives – like woodworking glue – epoxy 'cures' rather than 'dries'. Additionally, unlike woodworking glue, there is no shrinkage after it cures.

◉ MITRE JOINT

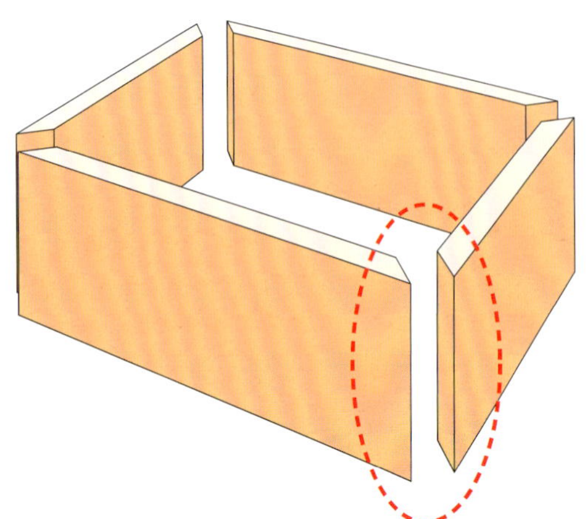

◉ STRUCTURAL DIAGRAM

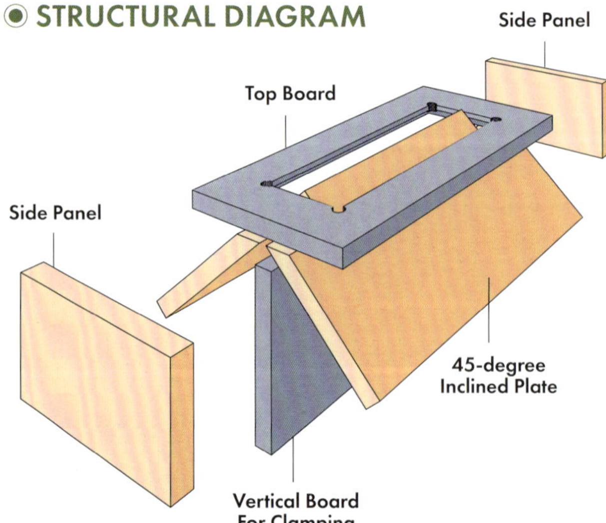

DIMENSIONAL DRAWING

Units: mm

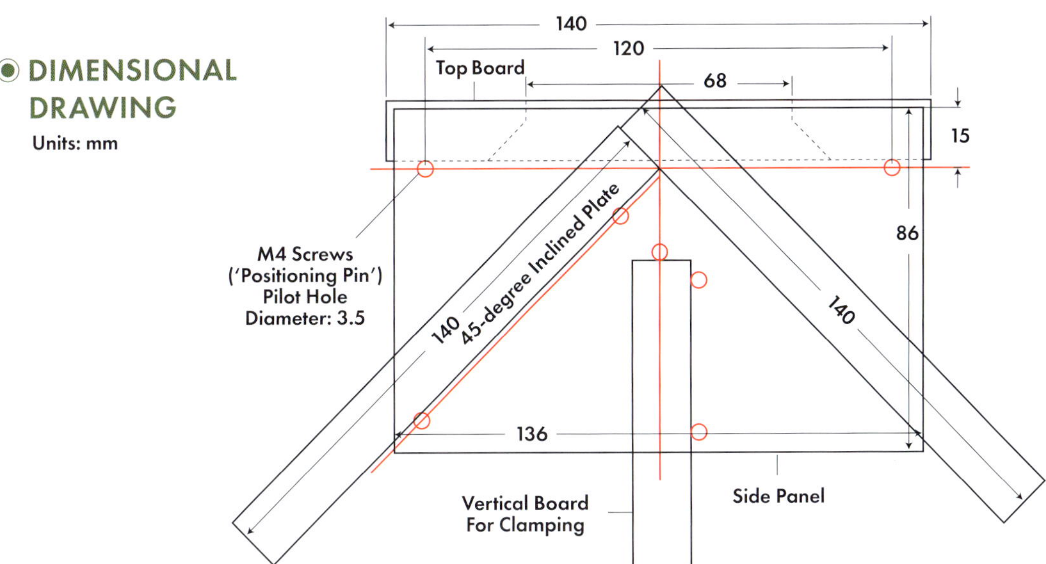

SIDE PANELS AND POSITIONING PINS

1 Join two side panels with double-sided tape and drill holes at specified positions (red circles). These are ⅛in (3.5mm) pilot holes for M4 screws. Before drilling, use a punch or awl to create a slight indentation at each hole position. That way the bit won't slip. Once the holes are drilled, insert M4 screws. Make sure they extend about 9⁄16in (15mm) out the opposite side (note the material above has a groove; it's scrap material and the groove is unrelated to the jig construction).

2 Both side panels have pins in the same position.

FIX POSITION USING PINS

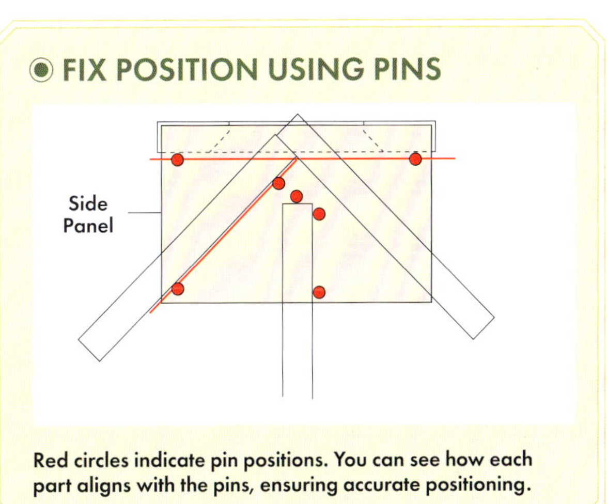

Red circles indicate pin positions. You can see how each part aligns with the pins, ensuring accurate positioning.

ATTACHING 45-DEGREE INCLINED PLATES

3 First, glue two boards together at right angles to create a 45-degree inclined plate. Use a square and a triangle to ensure a precise right angle.

4 To reinforce inclined plates, glue triangular pieces in the inner corners. Cut a rectangle out of the top board by pre-drilling four holes in the top board and then making straight cuts, from hole to hole, with a jigsaw. The opening should be sized so that base plate doesn't fall through.

5 After gluing, cut off both ends as shown. This will prevent interference with the top board. Use a hand saw to make cuts.

6 The top board underside should be chamfered significantly using a 45-degree chamfer bit.

◉ CROSS-SECTIONAL VIEW

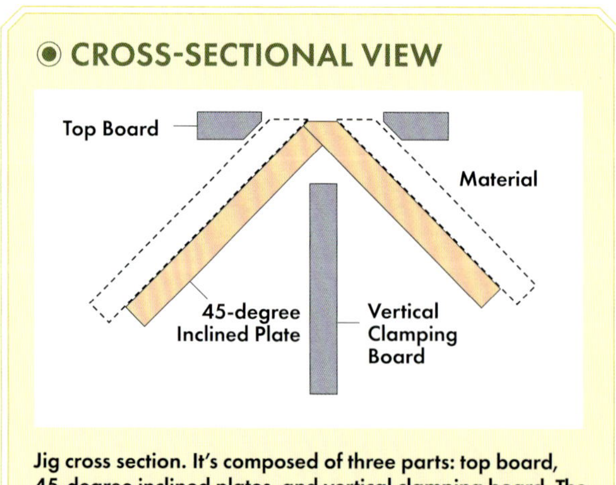

Jig cross section. It's composed of three parts: top board, 45-degree inclined plates, and vertical clamping board. The material to be processed (dotted line section) is clamped to the 45-degree inclined plates.

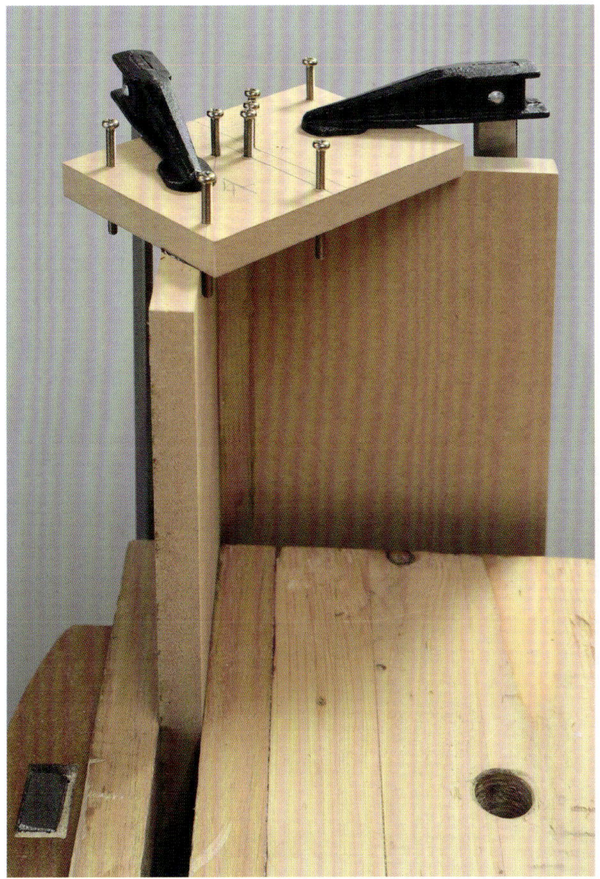

7 Gluing a side panel in place. Lightly clamp the plate to ensure the 45-degree inclined plate doesn't lose contact with the positioning pins. Once the epoxy has cured, glue on the other side plate. The clamping method for the marked area will be explained below.

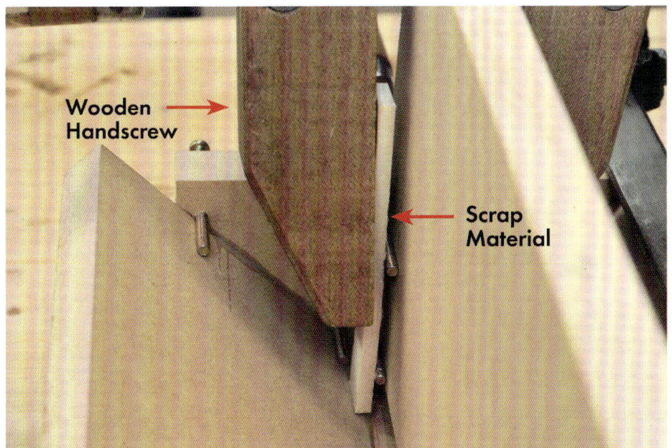

8 To prevent components from losing contact with pins, use scrap material to bridge two pins and clamp with a type of wooden clamp called a handscrew. It's used in situations where standard F-clamps or C-clamps are difficult to apply.

9 Attach the vertical clamping board.

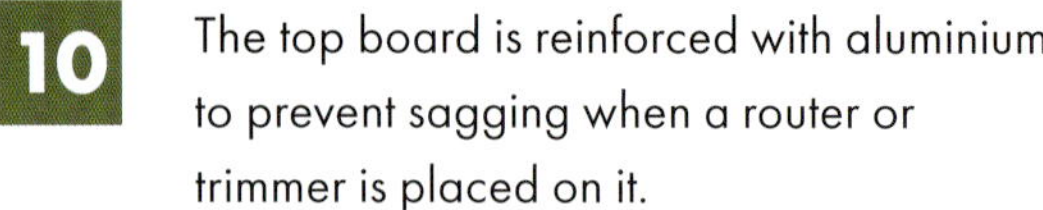

10 The top board is reinforced with aluminium to prevent sagging when a router or trimmer is placed on it.

11 Glue the top board to the main body of the jig. Place the support block on the top board. Lightly clamp the bottom edge of the side plate to the support block to stop it moving.

12

Cut off the top portion of the 45-degree inclined plates using a flush cut saw, and it's complete. The photo shows a replacement blade attached to a square block with double-sided tape.

13

Insert material from below while the router is mounted. Clamp it to the 45-degree inclined plate where it touches router's base plate. When using a trimmer, make a larger custom base plate (see page 44).

Height Gauge/Depth Gauge (Setting Trimmer Bit Cutting Depth)

The left item below is a height gauge made with a 6in (150mm) ruler. The right side slides up and down, allowing you to adjust to the necessary height. To set the height, place the trimmer on the tabletop and push it up until the bit touches the gauge. The right side features a combination square that can be used as a height gauge.

The height gauge ruler on the left is attached with an embedded magnet. The combination square on the right is a Starrett Double Square.

A hose clamp can be used as a depth stop. Cutting too deep all at once can put significant strain on the bit. To make deep cuts, start by cutting shallow and gradually increase the depth. When you reach the stopper position you have hit the desired depth.

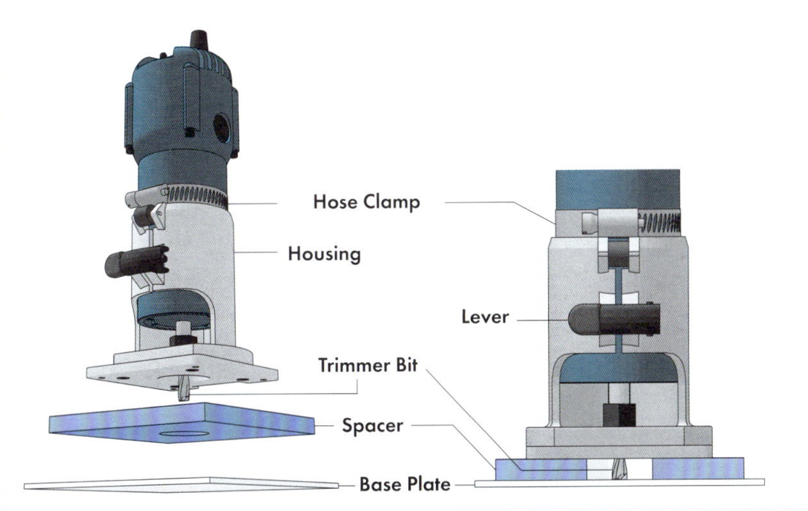

DEPTH SET PROCEDURE

1. Set base plate on workbench.
2. Place spacer on top of the base plate.
3. Place trimmer on top of the spacer and lower until it touches the base plate.
4. Lock the housing lever.
5. Tighten screw to secure hose clamp in contact with housing.

Easy to Make Mitre Joint Box 1

You can use a technique similar to folding cardboard to create a box. The only trimmer bit you'll need is a V-groove bit. This method has been featured in various YouTube videos.

Here, 11⁄32in (9mm) MDF is used. All four sides of the square-cut MDF are cut, with a straight guide (edge guide) attached to the trimmer, to create V-grooves. The grooves are cut very deep, leaving just a thin layer of the material connected together. By folding along those grooves, you can create a very simple mitre. It's crucial that the corners of the initial square are cut at precise right angles. The distance from the straight guide to the centre of the bit will determine the height of the box. There are limitations on the distance you can set, so check the maximum distance of your edge guide in advance.

PREPARATION

1 Here we have a straight guide attached to a trimmer. Adding a scrap piece as a fence makes it easier to use.

2 This is a 90-degree V-groove bit. Note that a larger bit is preferable as it can be used to process materials around $\frac{11}{32}$in (9mm) thick.

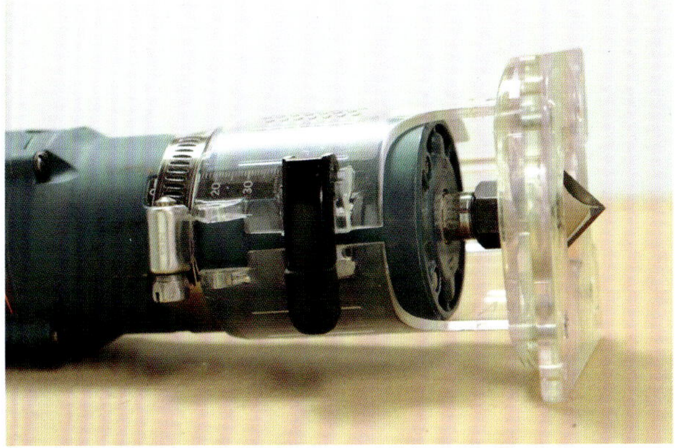

3 Combine acrylic spacers to set the bit depth to approximately the same thickness as the target material (see page 57). Once the depth is set, fasten the hose clamp stopper to avoid cutting too deep (see page 37). Cutting should be done in several passes to complete the V-groove.

V-GROOVE PROCESSING

4 Position the straight guide on the right, based on the direction of travel, and guide it along the material to create a groove. Make sure the bit isn't in contact with the material before turning on the trimmer, otherwise the trimmer could suddenly jump in any direction when switched on!

TIPS FOR STARTING AND FINISHING A CUT

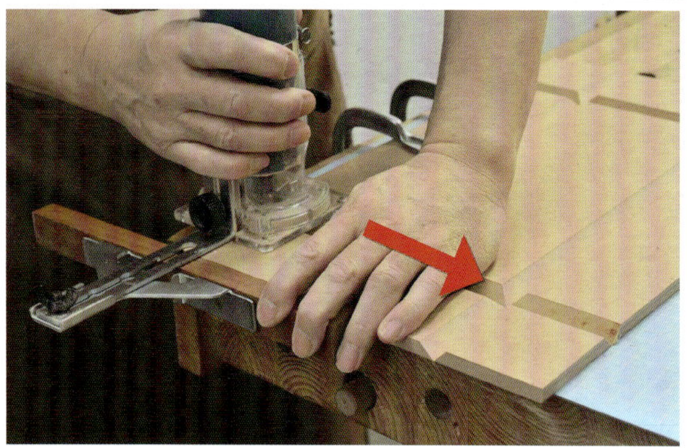

5 To start the cut, press the front part of the straight guide firmly against the material – as shown.

6 During the cut, apply even pressure across the entire straight guide. To finish the cut, slow down and keep the back of the straight guide pressed against the material until the cut is complete.

7 Gradually lower the bit. As you approach the maximum depth, flip the material over and apply masking tape to the back side of the groove. This way, if the material accidentally separates, the tape will help prevent it from completely coming apart.

8 The material will easily bend when cutting at the final depth – this is the desired outcome.

9 Cut four sections apart with a utility knife.

BONDING

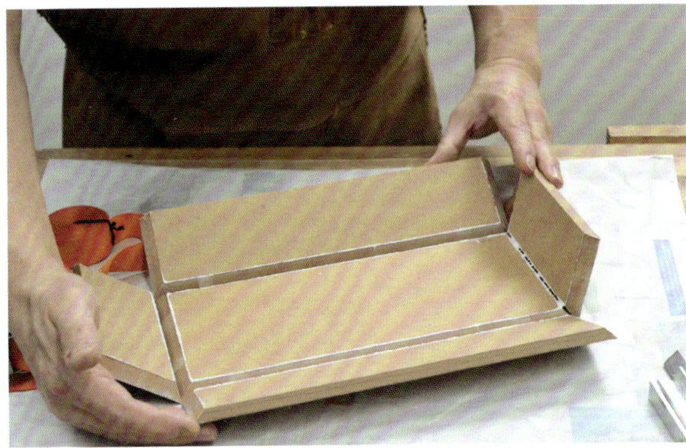

10 Fold the material and dry-fit to ensure there aren't any issues. Apply glue to all 45-degree edges and fold into a box shape.

11 Use the aluminium angle guide and a ratchet strap to tighten the assembly. Wipe off any excess glue with a damp paper towel.

12 Check corners for right angles using a square. If adjustments are needed to fine-tune the angle, just hook a string around the aluminium angle guide on the diagonal and tighten.

13 Finish using sandpaper or a plane, then it's complete!

ADDING REINFORCEMENT

14 Our mitre joint is only glued on the end grain, so passing toothpicks or bamboo skewers through the joint is a good method of reinforcement. Once the epoxy has cured, cut off excess with a flush cut saw and finish with a plane. For details on reinforcement, see 'Mitre Joint Reinforcement Drill Guide' in the diagram below.

◉ MITRE JOINT REINFORCEMENT DRILL GUIDE

This guide will help you drill pilot holes in the mitre joint and insert toothpicks or bamboo skewers for reinforcement.

Mitre Joint Reinforcement Drill Guide Dimensional Diagram (9mm Board)

Units: mm

Base
Drawing a Pencil Line
30
9
20
Guide Block
HORIZONTAL PROJECTION

30
Hole for Toothpick/ Skewer
FRONTAL PROJECTION

Black Line
Guide Block
Base (Size as Necessary)
Guide Hole
ISOMETRIC VIEW

The 45-degree cuts are made with a 'Saw Guide F' (see page 47). The adhesive used here is woodworking glue. Guide holes should be drilled beforehand with a drill stand or a drill press.

Size adjustments are easy! This makes it an excellent project for a handheld trimmer. Give it a try!

Fix guide blocks to the box corners and drill holes into the box. This guide block is designed so that the pilot hole passes through the centre of the mitre joint 45-degree edge at 11⁄32in (9mm) thick. The guide hole positions will change depending on board thickness.

In this case, the guide block is made from hard maple scrap. The base material size is flexible. Use a square to ensure the guide block is attached to the base at a right angle. The photo above on the left shows spring clamps and pliers that are used for gluing 45-degree edges. Please also note the photo above right: It shows spring clamps and pliers being used for making frames.

3

Chapter 3

Trimmer Table

Requirements for a Trimmer Table

A trimmer table involves mounting a trimmer under a flat surface (the tabletop). This allows just the bit to protrude above the surface. A straight edge, called a fence, is used to guide the workpiece along the bit for processing. The fence serves as a tool to determine the positioning of the workpiece during machining.

Using a trimmer table not only expands the range of processing you can perform, it also improves accuracy. In handheld processing, the trimmer can become unstable, whereas, on a trimmer table, both the trimmer and the material remain stable. When grooves need to be cut across the material (like the end grain or edge grain), a trimmer table is seriously advantageous. High processing accuracy that creates professional-grade joints is the main reason for using a trimmer table.

You can create a clean and quiet working environment by employing a vacuum cleaner for dust collection, a speed controller for the motor, and a box-type stand beneath the tabletop. This setup allows for woodworking almost anywhere, even in places like apartments where trimmer operations might be challenging.

In Chapter 3, the author will first introduce the essential conditions for a truly useful trimmer table.

TABLETOP

Whether using a commercially available trimmer table or building your own, a perfectly flat tabletop and straight fence are essential.

The most important component of a trimmer table is the tabletop. If it isn't extremely flat, precise work is nearly impossible. Additionally, the table should be as thin as possible. A thicker tabletop limits the height of the protruding trimmer bit. However, it must also possess enough strength to prevent warping during use, even if it's thin. While it may seem contradictory, the ideal tabletop is flat, sturdy, and thin. When building your own, consider using materials like acrylic or other plastics for the tabletop. Reinforce from below with L-shaped angle brackets to prevent warping. For commercially available products, choose a metal tabletop that is designed to be warp-resistant from the outset.

The tabletop size can vary depending on what you intend to make. It's flexible. For larger projects, such as cabinets, dressers, or tables, a larger trimmer and trimmer table is recommended. Still, it's best to use the trimmer table for projects that aren't too large. In this book, the tabletop is a commercially available 16×16in (400×400mm) MIRAI Trimmer Table Set.

BIT OPENING

There's another issue to address with a trimmer table: how to manage bit openings for various bit diameters. Smaller bits should be used with smaller bit openings. Using a smaller bit with a larger hole creates a gap that can cause materials to catch. This can be quite dangerous. To solve this, many commercially available tables feature removable insert rings that allow you to adjust the hole size to match the bit being used (see diagram below left). While this might seem convenient at first glance, an issue arises from the slight step between the rings. When processing material, it can catch on the step, thus jamming the material. It's unexpected and dangerous for material to catch and stop while the bit is spinning at high speeds. Additionally, this problem can reduce processing accuracy.

As a trimmer table beginner, you may not notice the issue with jamming material much, but as you improve and seek higher precision, differences in height will gradually become more noticeable. When making your own table, you should avoid incorporating structures that are prone to such height differences from the start. One solution for a DIY table is to place a smaller board with a smaller bit opening on top of the main table, as shown in the illustration above.

The 'MIRAI Trimmer Table Set' consists of an aluminium top and a sliding base component. It features a zero-height top without insert rings, with two bit openings of different sizes instead. The user can switch between them simply by sliding the trimmer. This book will explain how to create a DIY trimmer table using this product.

MITRE GAUGE SLOT

Some commercially available tables have grooves on their tops (red grooved area). This is called a mitre gauge slot. A mitre gauge, shaped like a protractor, is fitted into this slot and slides along to primarily assist with end grain machining. However, it's often not very useful, and in DIY versions, it only complicates the construction of the tabletop. The author believes it is unnecessary.

FENCE

The fence is used as a straight edge to fix the position of the workpiece. By adjusting the distance between the trimmer bit and fence, you can easily make rabbet cuts and grooves. The fence is securely fixed to the tabletop with clamps or similar devices.

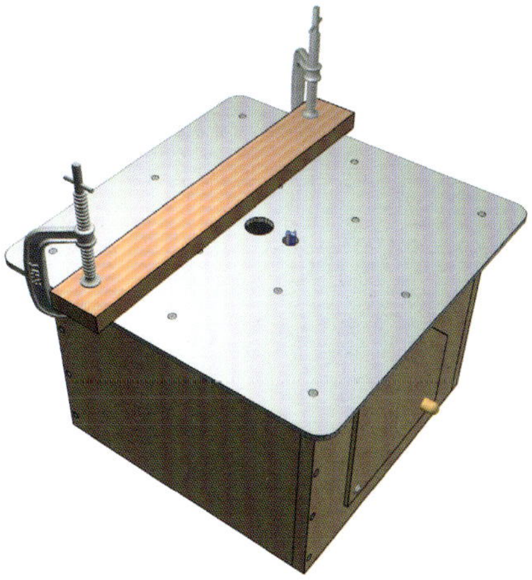

1

A simple straight edge fence. Used for cutting grooves.

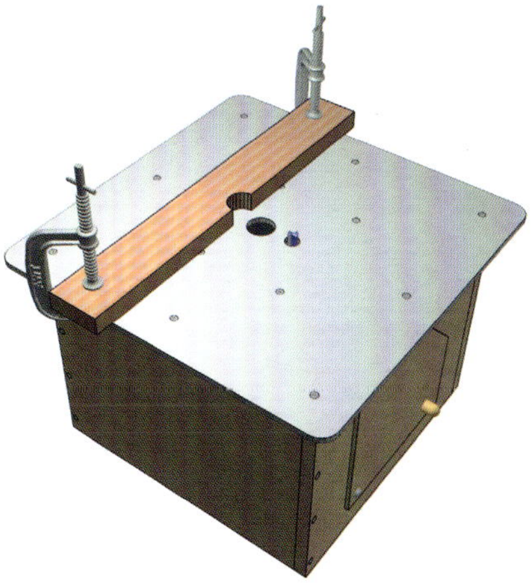

2

This fence has a groove where the trimmer bit can fit. You can perform rabbet cuts by adjusting the amount the bit protrudes from the fence surface.

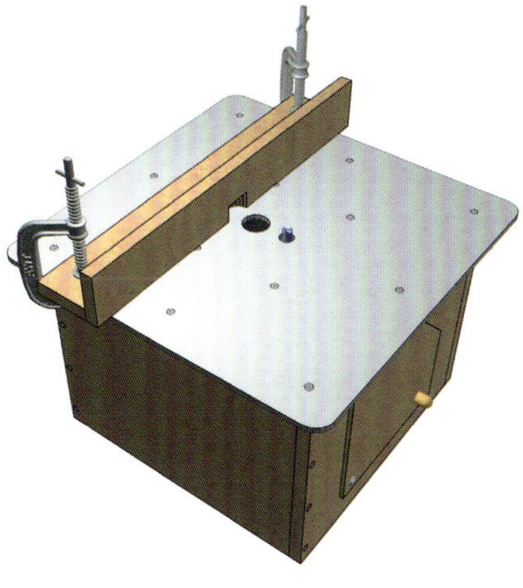

3

This fence's cross-section is L-shaped. It allows you to place target material upright against the fence so you can make rabbet cuts and grooves.

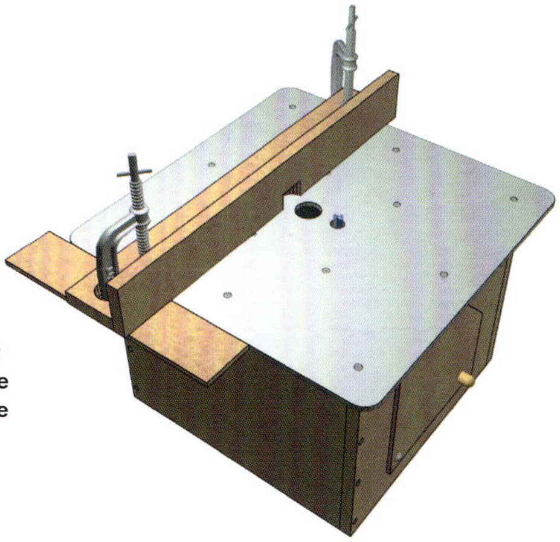

4

This fence is T-square shaped. This shape makes it easy to adjust the material length cut by sliding the target piece along the fence.

Fence Dust Collector and Box Dust Collector

Efficiency increases when a dust collection port is added to the fence. When rabbet cutting, the shavings are vacuumed right up, leaving almost nothing on the tabletop. This prevents obstruction when conducting your next operation and is very effective for continuous work.

A T-square-type fence, with parallel plates attached at the ends, can be moved back and forth on the tabletop, just like a drafting T-square. When attaching the main body of the fence and the short T-square plate to your DIY project, insert a 0.5 to 1mm thick piece of cardboard, or plastic sheeting, between the two parts (see illustration). The parallel plate will be slightly lower than the tabletop surface and prevent material from catching if it extends past the end of the fence.

T-SQUARE FENCE

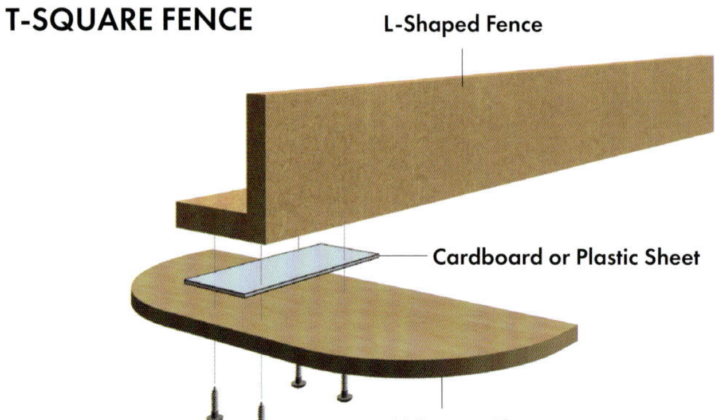

Trimmer table with dust collection hose and speed controller attached. It's being used for cutting grooves, so the hose is connected to the box stand and not the fence.

The two diagrams below illustrate the difference between fence and box stand dust collection. Dust collection varies depending on the part of the material being processed and the method of processing.

RABBET CUTTING DUST COLLECTION

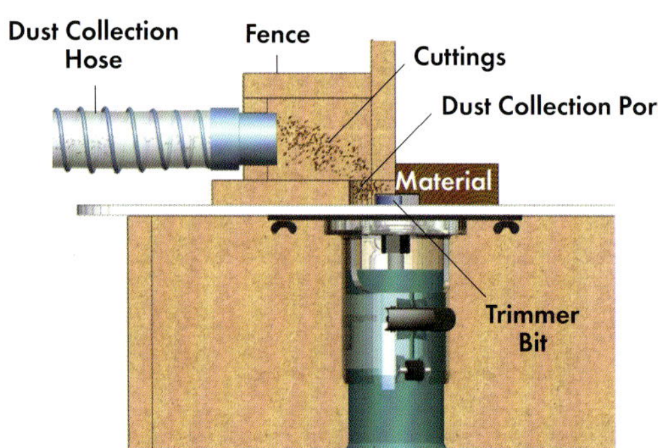

The cutting area of the material faces the dust collection port when rabbet cutting. This allows dust to be collected from the fence.

GROOVE CUTTING DUST COLLECTION

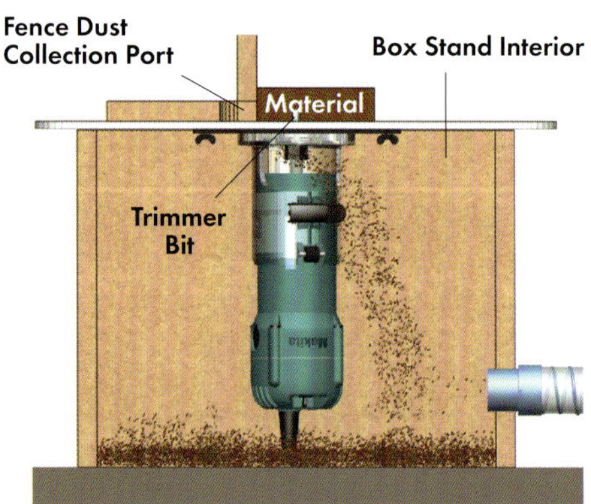

Dust cannot be collected through the fence when groove cutting. Instead, a dust collection hose is connected directly to the box stand. Dust is collected through the tabletop bit opening. Luckily, shavings hardly accumulate in the vacuum cleaner, instead they tend to collect at the bottom of the box.

Fence With Dust Collection Port

Two different types of fences with dust collection ports. Please use them as references for your own DIY projects.

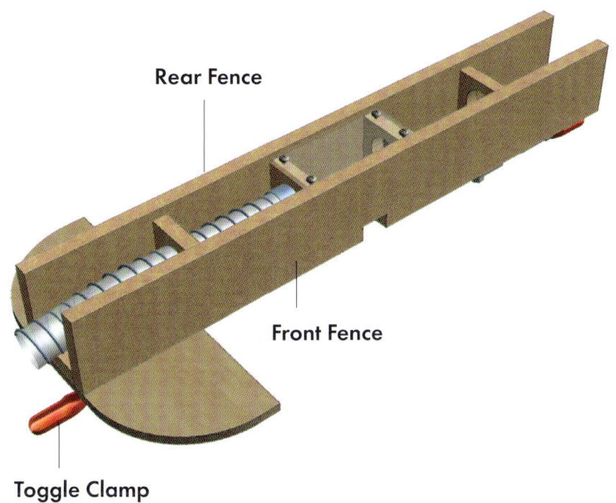

This fence features a dust collection hose that runs parallel to the fence itself. Note that the backside can also be used as a fence and, since there is no dust collection port on that side, there is no risk of material catching on the dust collection port. This fence is fixed to the tabletop using two toggle clamps.

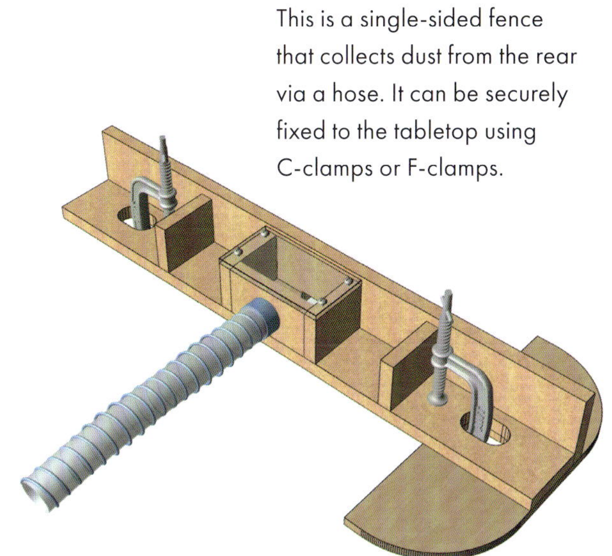

This is a single-sided fence that collects dust from the rear via a hose. It can be securely fixed to the tabletop using C-clamps or F-clamps.

USING SPACERS WHEN MOVING THE FENCE

When cutting grooves, the width of one pass is determined by the width of the bit. However, by moving the fence and performing a second cut, you can widen the groove. The method for moving the fence is crucial. There are two approaches. The first method is suitable for cases where precision is not a primary concern. You can use a T-square type fence and mark the initial position on the tabletop with a pencil. Then, measure the distance required to widen the groove and mark the new position to set the fence. Reading the ruler to determine the distance for adjustment relies on estimation, making it less than ideal for precision cutting. The second method involves using spacers to achieve far more accurate adjustments. You should prepare spacers of varying thicknesses and then just combine them as needed. Details of this method will be explained in the following pages.

When widening grooves, one important note is that, for both methods mentioned earlier, the fence should be moved away from the router bit. If the fence is moved closer you will be making a climb cut, which is very dangerous (see page 82, 'Rules for Widening Groove Widths').

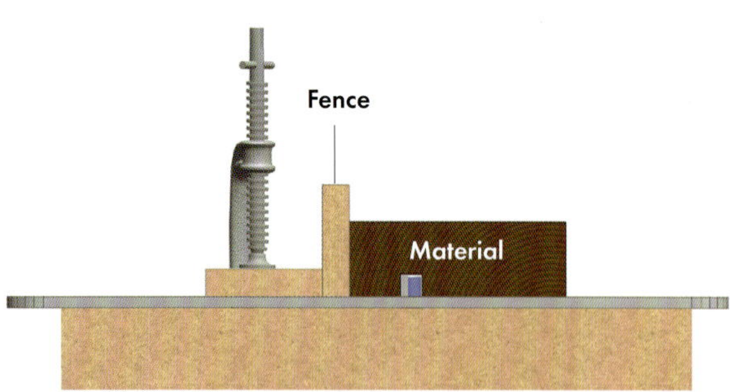

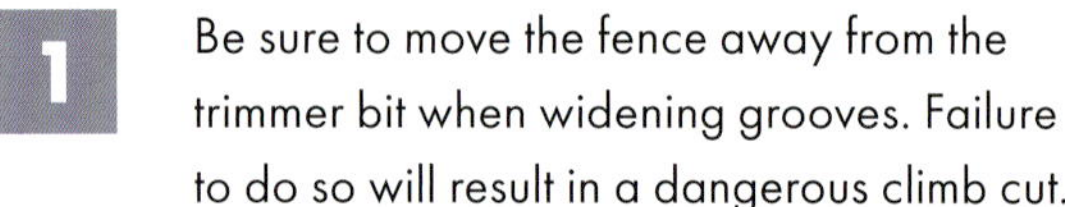

1 Be sure to move the fence away from the trimmer bit when widening grooves. Failure to do so will result in a dangerous climb cut.

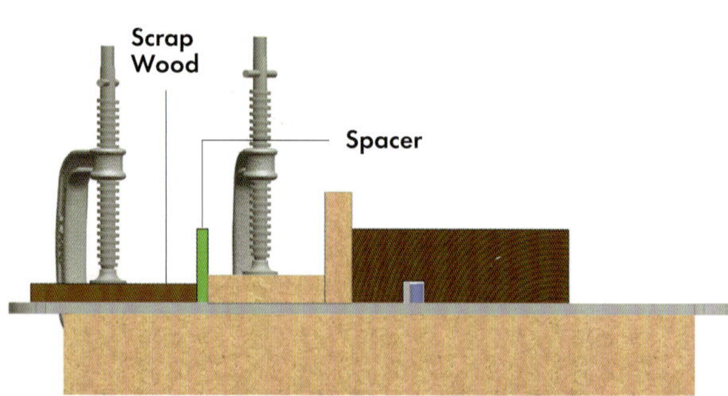

2 Place the spacer against the fence back and secure with scrap wood. The spacer thickness will determine the distance the fence is moved.

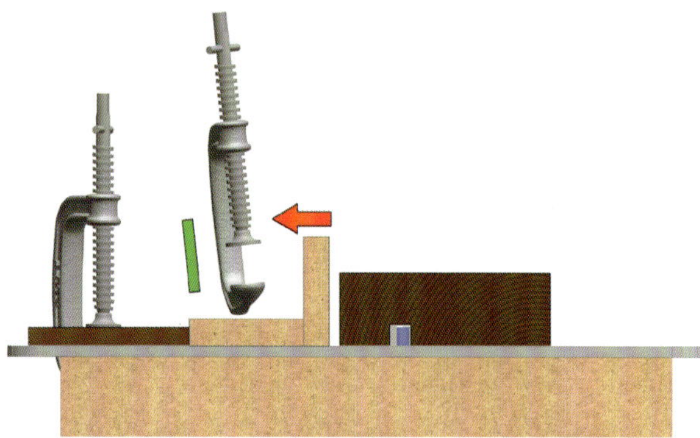

3 Remove the clamp and take out the spacer. Position the fence against the scrap wood. This completes the fence adjustment.

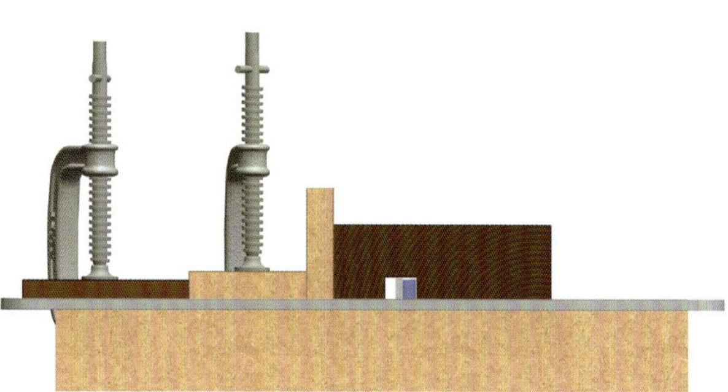

4 Clamp the fence in place. Press the material against the fence and make a second cut to widen the groove.

FENCE HEIGHT

Sometimes you need to process material while it's standing upright against the fence. The fence needs to be high enough to allow for the proper application of pressure. If the fence is too low, your hands might be too close to the trimmer bit. This is quite dangerous. In such cases, it's best to set the height at a minimum of 3⅛in (80mm) or more. Additionally, make sure that the fence surface is perpendicular to the tabletop.

THE BOX STAND UNDER THE TABLETOP

For a basic stand, you can simply attach four legs. However, here, I will introduce a box-type stand designed to minimize noise. This stand features soundproofing sheets and includes an access door. When used in combination with a speed controller, you can significantly reduce noise levels.

When determining the stand size, make sure the tabletop extends at least 1⅛in (30mm) past the edge on one side. This creates a clamping space, making it easier to secure the fence and other components to the table.

NOISE REDUCTION METHODS

Three Methods for Noise Reduction

1. **Make a box-type stand to contain the noise.**
2. **Apply noise reduction sheets inside the box.**
3. **Use a speed controller. Be careful though as some speed controllers can't be used with trimmers that have a soft start feature. Confirm trimmer options when purchasing.**

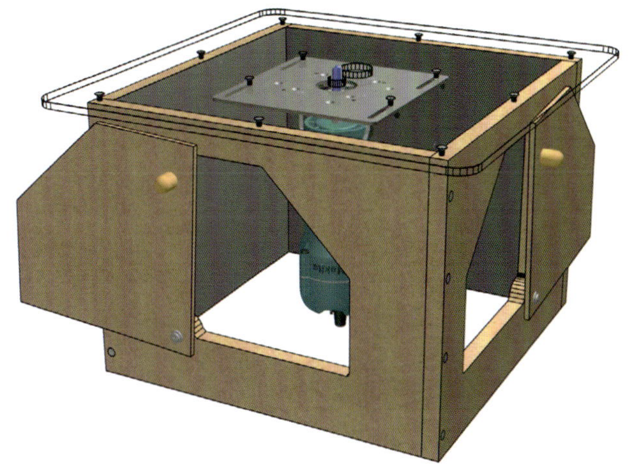

DETERMINING THE STAND HEIGHT

You can create a user-friendly trimmer table if you ensure that the tip of the straight bit can be lowered to the level of the tabletop.

To change the bit, tilt the entire trimmer table 90-degrees to one side and pull out the motor. This box stand has no bottom panel, so any shavings sucked inside through the bit hole accumulate directly on the work surface.

STAND HEIGHT

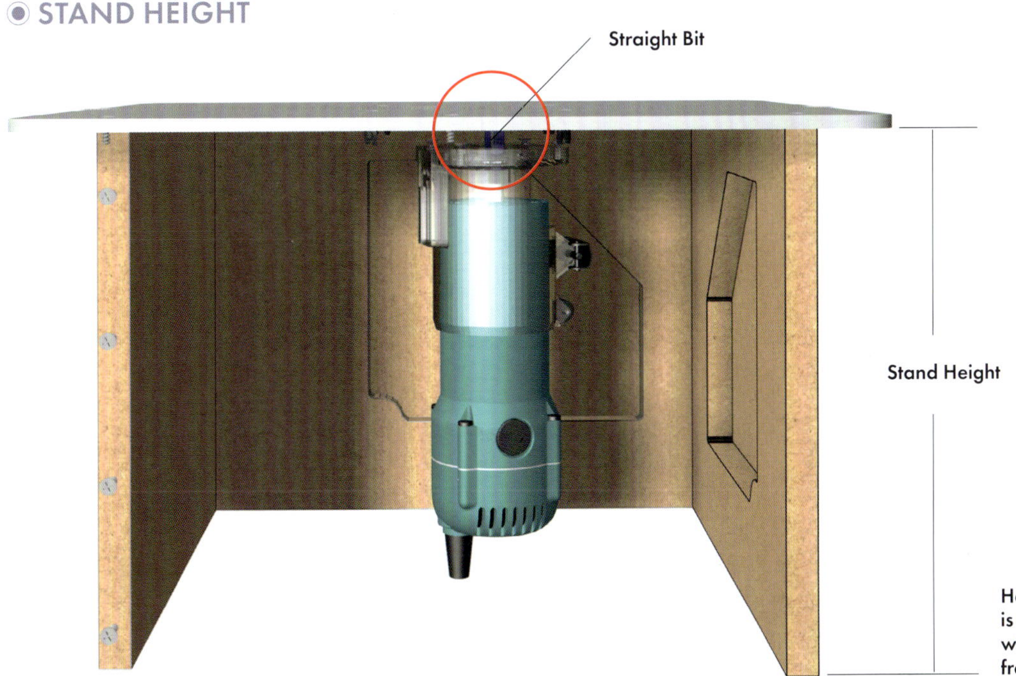

Trimmer Table Usage

On a trimmer table, the bit rotates anticlockwise, i.e. it rotates in the opposite direction of handheld processing. Therefore, the feed direction for a normal cut is from right to left (see page 22). It's important to remember this.

As shown in the diagram on the next page, in a normal cut, the cutting edge of the bit rotates into the end grain. This is correct. Part of the diagram replaces the trimmer bit with a tyre. You can see that the tyre's direction of rotation and the movement of the material collide, resulting in push back. This is normal.

On the other hand, in a climb cut, material is fed from left to right. The cutting edge of the bit doesn't strike the end grain of the material first. Instead, it strikes the face of the wood. This causes the cutting edge to catch and pulls the material – in the direction of the arrow – propelling it away with great force. Your hand can be pulled in the same direction, increasing the risk of coming into contact with the bit. This is extremely dangerous! If we replace the bit with a tyre, it becomes clear that the tire is pushing the material outward. Climb cuts are extremely dangerous and should never be attempted. Additionally, always wear safety glasses and ear protection while working.

NORMAL CUT

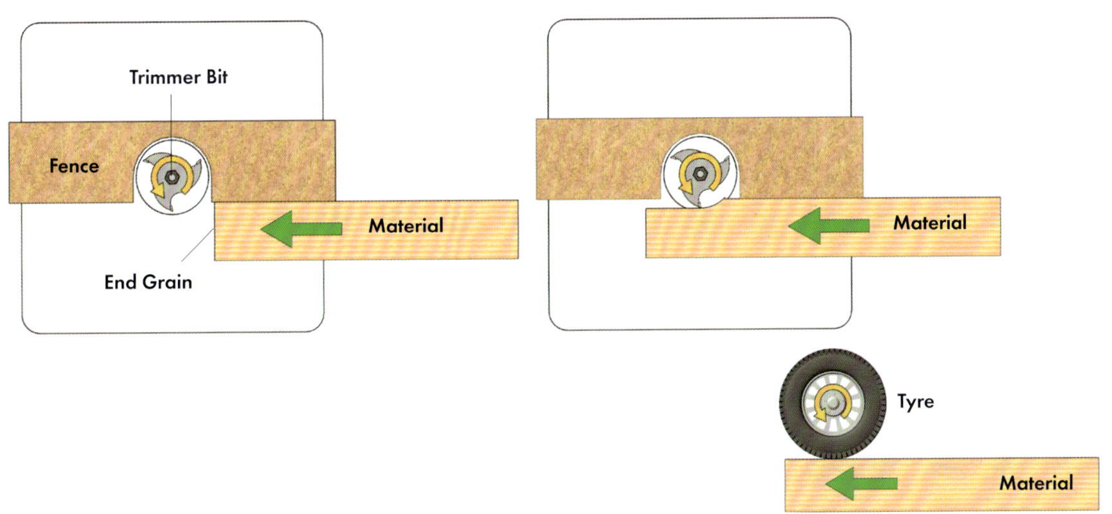

CLIMB CUT

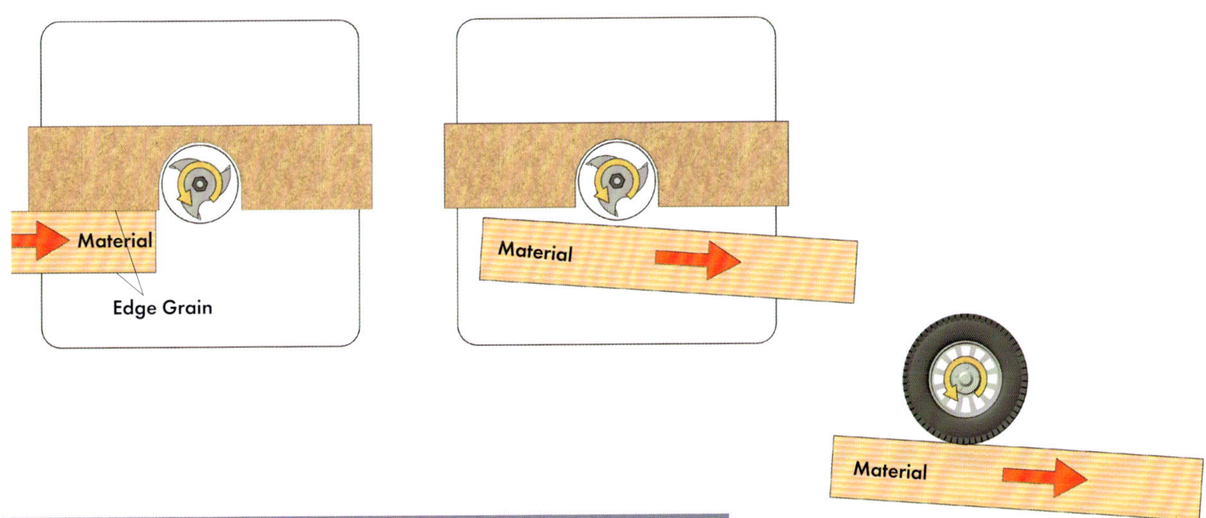

MATERIAL IS FED FROM RIGHT TO LEFT

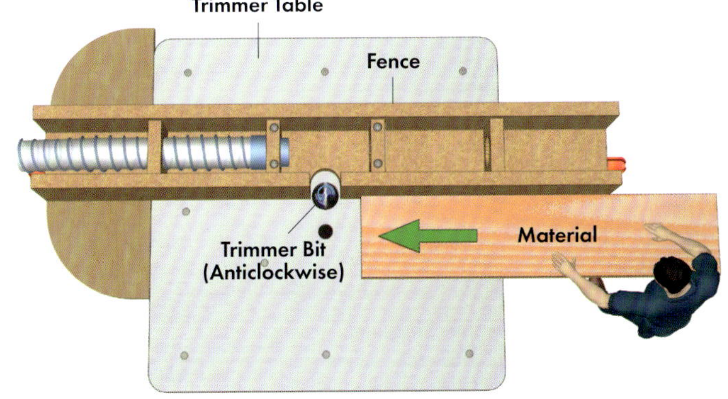

This photo shows the process of cutting a V-groove. The material is positioned against the fence and fed from right to left.

DANGEROUS ACTIONS

All three actions illustrated in the following diagrams are extremely dangerous. Please do not attempt them under any circumstances.

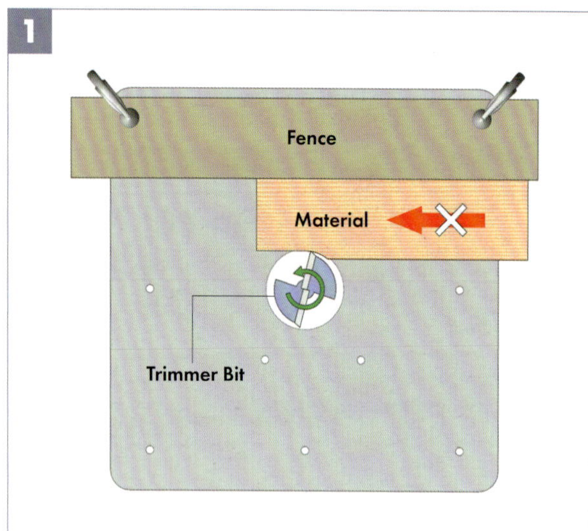

Avoid placing material between the fence and router bit. Diagram 1 depicts an extremely dangerous climb cut.

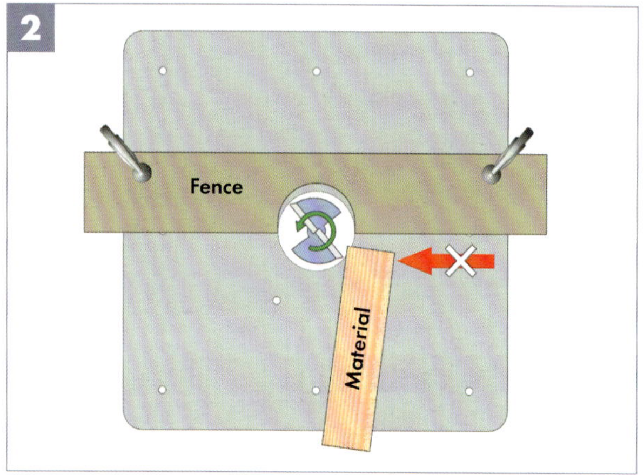

Short workpieces can become quite unstable, so caution is required. A jig to stabilize the material is almost always necessary.

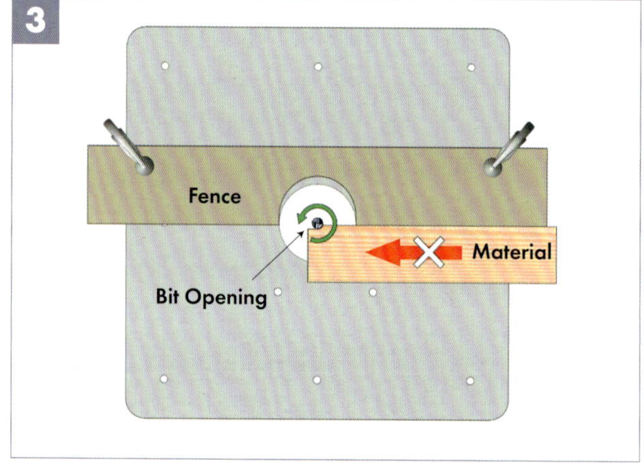

Always avoid using a thin bit in a large bit opening. If the gap is too large, material can catch, producing a dangerous situation.

BASIC OPERATION – RABBETING AND GROOVE CUTTING

Rabbeting and groove cutting are trimmer table basics. It's not an exaggeration to say that by mastering these methods, you can achieve nearly any type of woodworking. As shown in the diagram, even when using the same router bit, whether the action is considered rabbeting or groove cutting is determined by the distance between the fence and bit. Various other examples will be introduced on the next page.

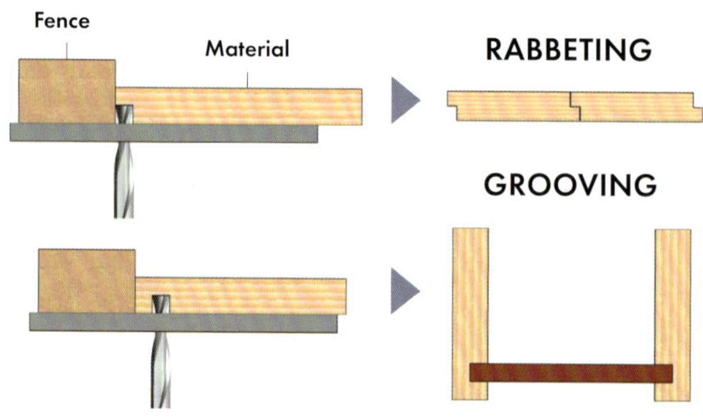

TRIMMER TABLE EXAMPLES

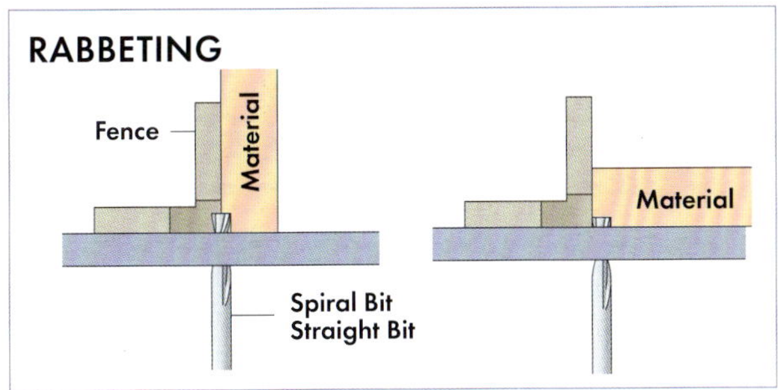

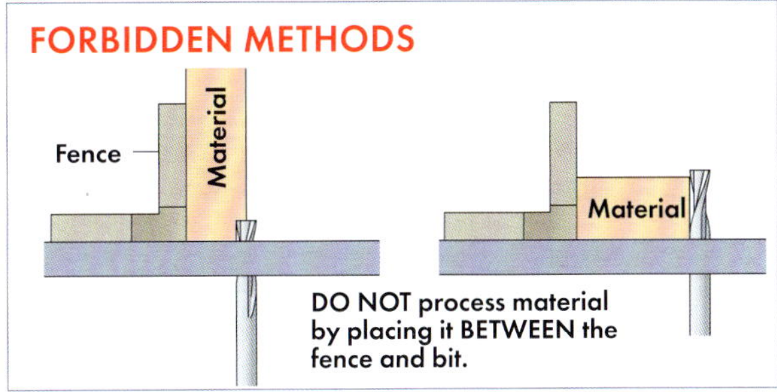

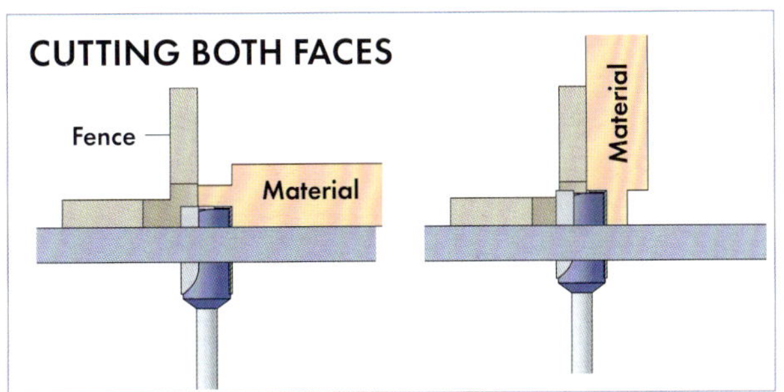

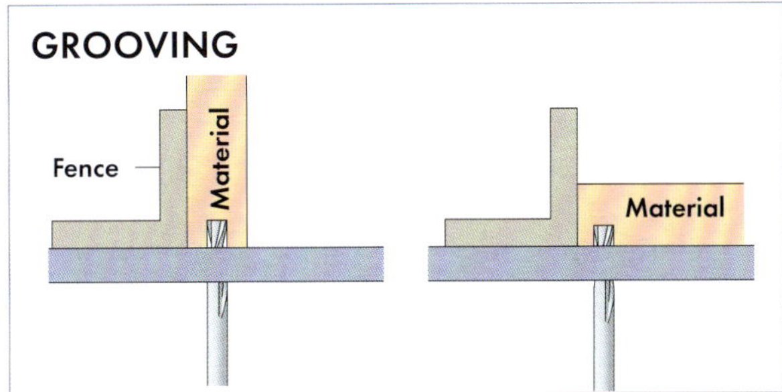

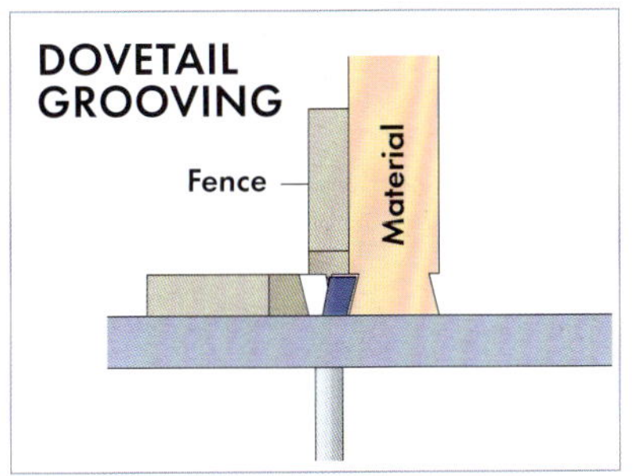

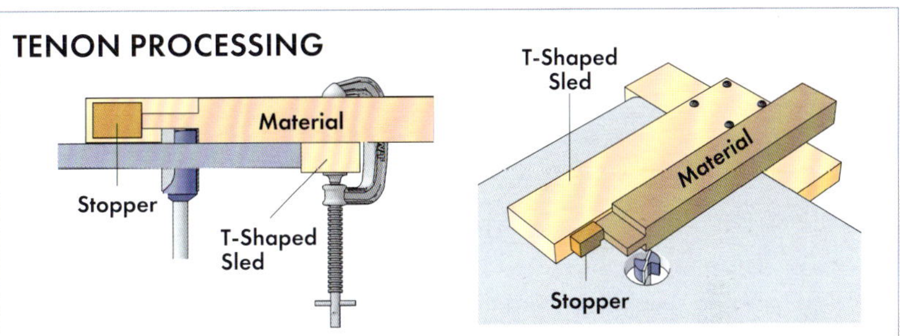

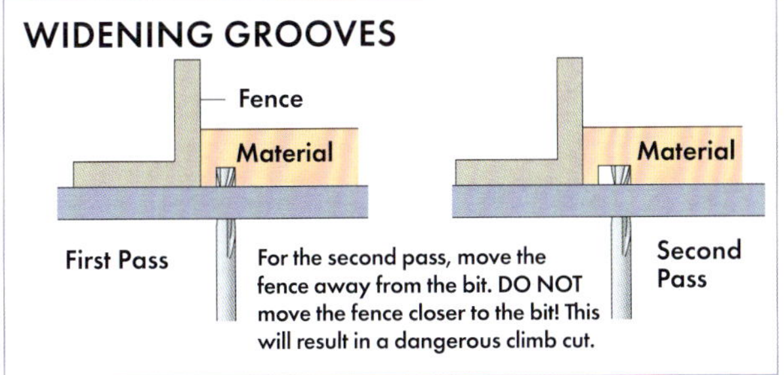

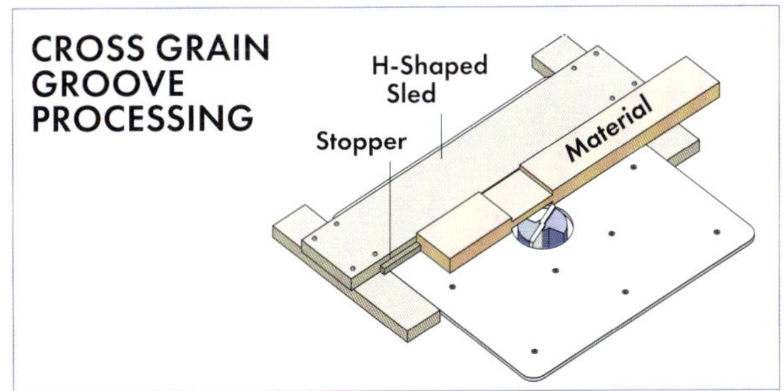

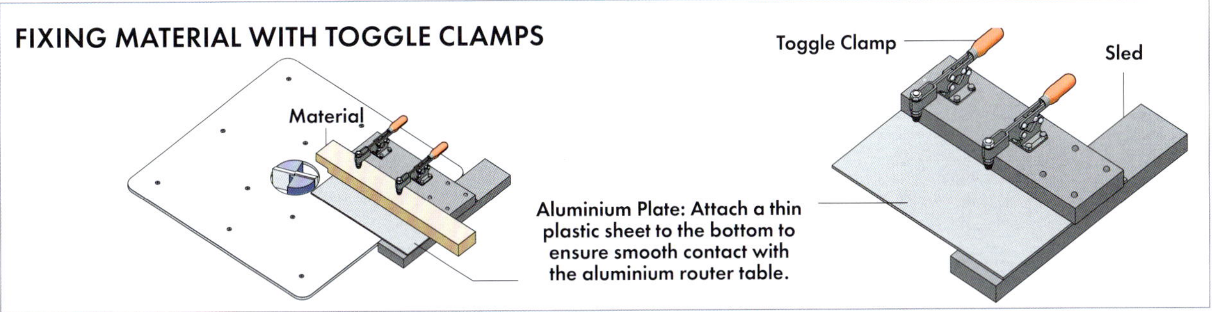

EDGE JOINTING WITH A TRIMMER

Here is a convenient method for creating straight and square finished edges. Use double-sided tape to temporarily attach a shim (spacer) – about 0.5 to 1mm thick – on the left side of the fence. Then, simply align the outer cut radius of the router bit with the shim to complete setup. After that, just feed material from right to left to machine the edge.

Secure the acrylic shim to the fence using double-sided tape. Adjust the fence, or use a ruler, to align the router bit tip with the shim surface.

Feed material from right to left to create a straight edge.

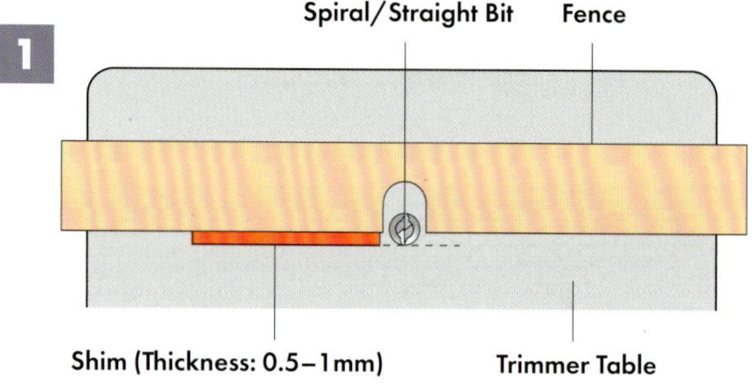

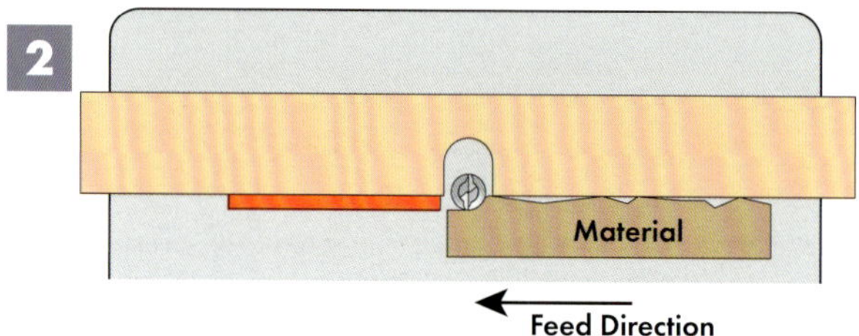

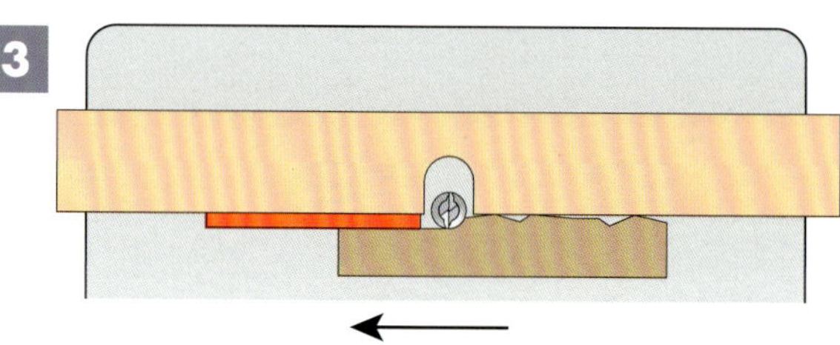

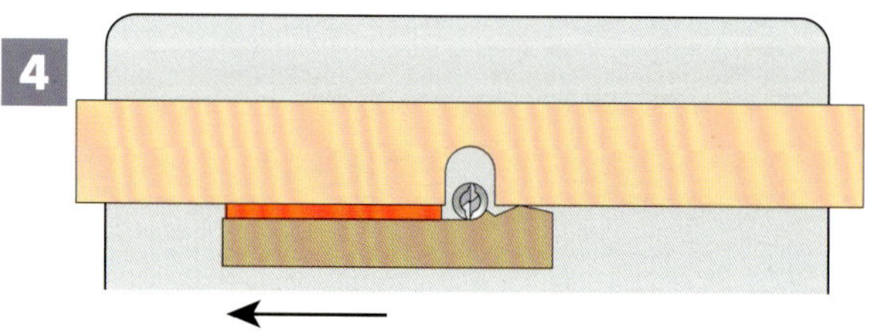

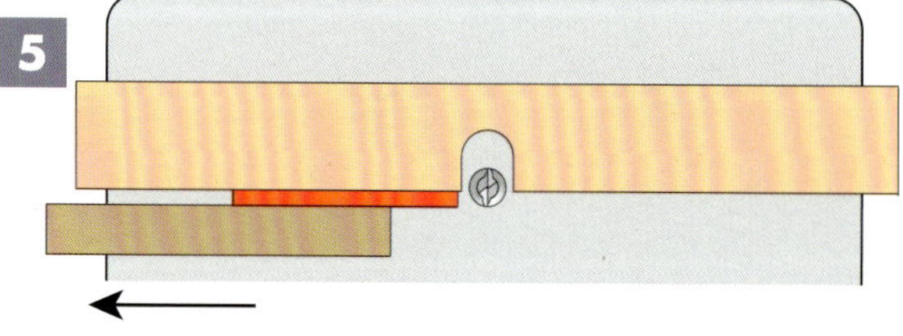

SETTING BIT AND FENCE GAP

This method sets the distance between the router bit and fence using a spacer. In this instance a metal rod is used in lieu of a bit. This can be quite convenient when using a straight bit that has the same shaft and cutter diameter. The 'silver' rod is ¼in (6mm) aluminium, while the 'gold' rod is ¼in (6.35mm) brass.

When using a straight bit, the edge of the bit must be perfectly perpendicular to the spacer to maintain accurate distancing. With a round rod, it's easy to set this distance because the entire circumference can be considered the bit edge. The O-ring on the brass rod serves as a depth stop when inserting it into the router's collet chuck.

This photo shows a spiral bit with its helical design. Since the cutting edge always contacts the spacer, regardless of its position, setting it up is easier than the two-flute straight bit – especially when a round rod isn't available!

Groove Widening

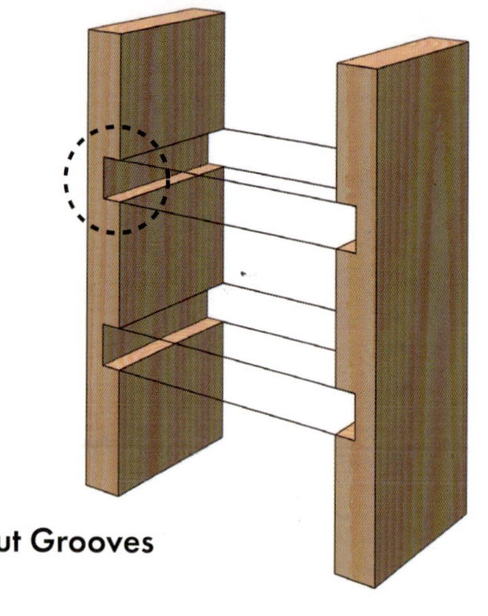
Cross Cut Grooves

Grooving Along the Grain

Grooving normally occurs in two ways. The first is to cut across the grain. These cuts are most commonly used for bookshelf joints, like the dado joint. The second is to cut along the grain. This cut is used to create pockets for items like the bottom panel of a box. Here, we will introduce a method for widening such grain-wise grooves using a trimmer table.

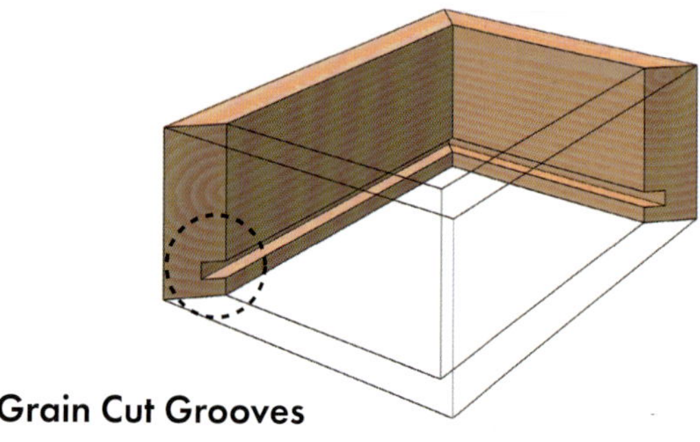
Grain Cut Grooves

Groove-widening techniques are necessary when you need groove widths that cannot be obtained with just a single pass of a spiral or straight bit. There are various methods available, but here I will introduce the method I use most often. To widen the groove in two passes, you obviously begin by completing the first cut.

Initially, we need to process the first groove. Then, by shifting the fence, you can perform the second pass to widen the groove. The amount (distance) you shift the fence is determined by spacers.

The procedure involves subtracting the cutting diameter of the bit from the thickness of the board that needs to fit into the groove (board thickness – bit diameter). The resulting value is the distance to shift the fence. This distance can best be achieved by combining spacers.

In in the bottom right photo, the 11/32in (9mm) piece of plywood fits perfectly into the groove after processing. In this case, the plywood thickness was first measured accurately with callipers. They showed 23/64in (9.2mm). A 1/4in (6.35mm) spiral bit is used, and by performing two passes, the groove width was expanded to 23/64in (9.2mm).

Various spacer thicknesses are shown. The length of each spacer should be around 12–16in (300–400mm). They can be purchased at most model shops. You can use the overhead projector sheets (OHP) for 0.1mm spacers.

PROCESSING STEPS

First groove being cut. As you can see, the fence has no opening on this side. When cutting grooves, the dust collection function can't be used regardless (see page 70), so we chose the side with no opening to prevent snagging. Note that, the box section of the router table stand has a drilled hole connected to a dust collector (vacuum). Dust can therefore still be collected through the bit hole. Here, a ¼in (6.35mm) spiral bit is used to create the groove. Next, we will prepare to widen the groove.

2

Photo shows rear of back. The difference between plywood thickness and bit diameter is 7⁄64in (2.8mm). That becomes the distance we need to move the fence. When the groove width is exactly that of the board thickness, it can sometimes be too tight. So, the groove was widened by an extra 0.2mm to make an even ⅛in (3mm). A ⅛in (3mm) acrylic sheet is used as a spacer and is placed between the fence and backup board, and that board is clamped to the table surface.

3

Next, remove the spacer, release the fence, and place it directly against the backup board before securing it again. This completes the ⅛in (3mm) movement of the fence. The important point here is that the fence was moved away from the router bit when it was adjusted (see pages 71 and 82).

4

Once the second groove is cut, the task is complete.

◉ RULES FOR WIDENING GROOVE WIDTHS – MOVE THE FENCE AWAY FOR THE SECOND PASS

When widening a groove, always move the fence away from the bit. If you move it closer, you will be making a climb cut (see pages 21 and 75). This technique appears frequently in this book, so please keep it in mind.

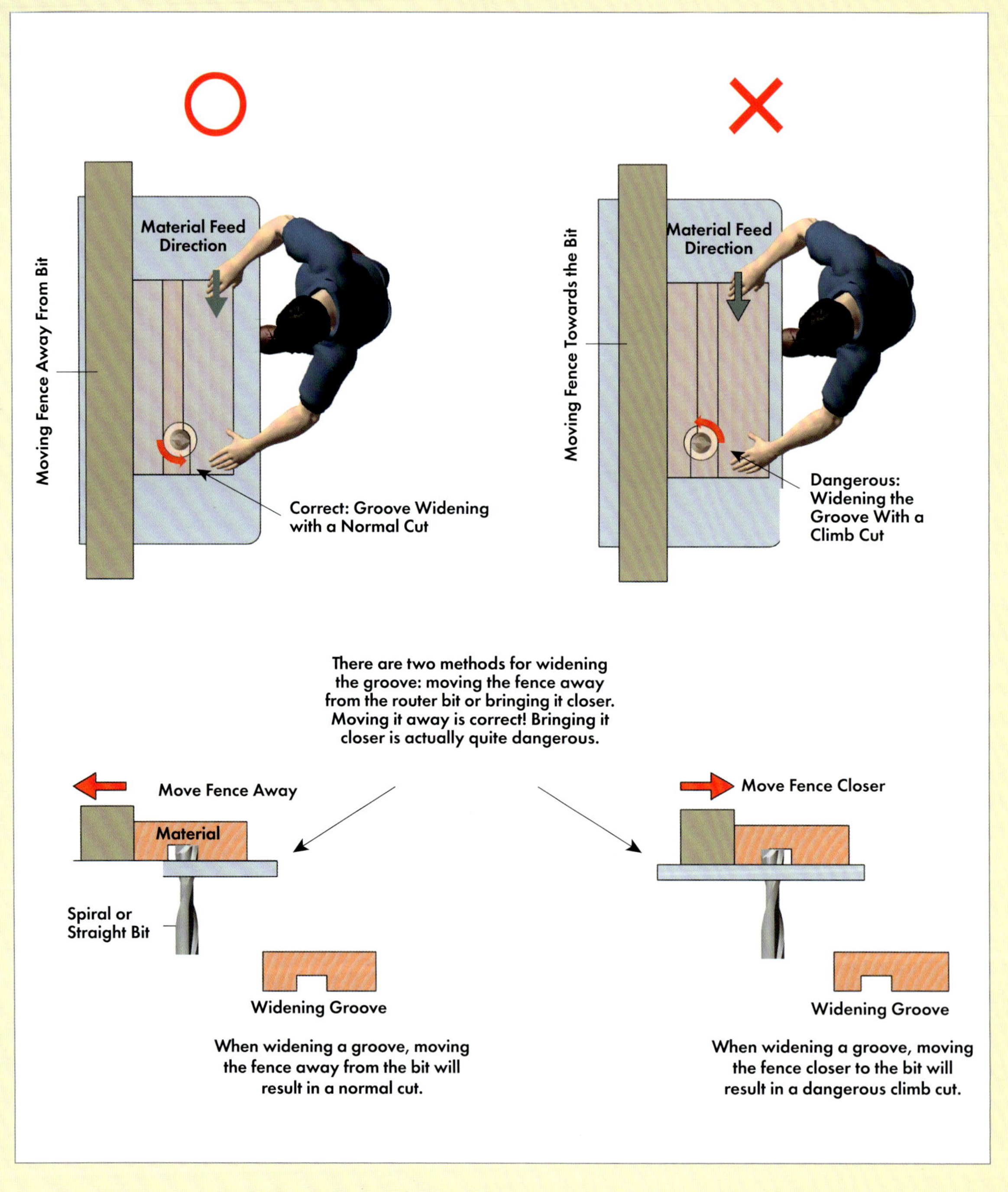

GRADUALLY REMOVING MATERIAL

When performing groove or rabbet cuts, be sure to remove material in increments that don't burden the bit. If the groove is deep, it's best to make cuts in multiple passes. For dovetail grooves, where the shape can't be easily cut, you might need to first use a different bit for rough cutting before cleanly finishing the cut. This approach helps reduce the strain on the bit and extends the lifespan of your tools.

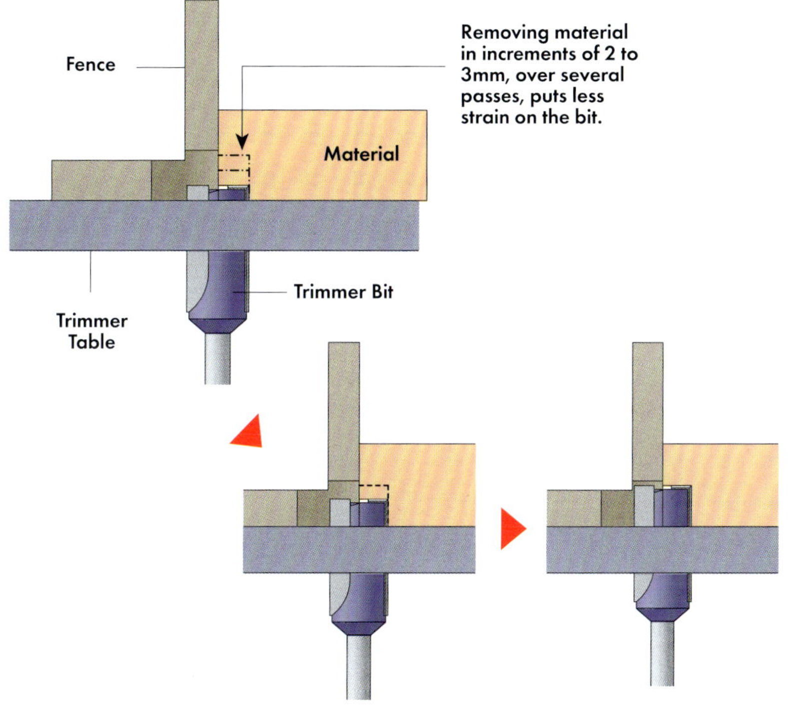

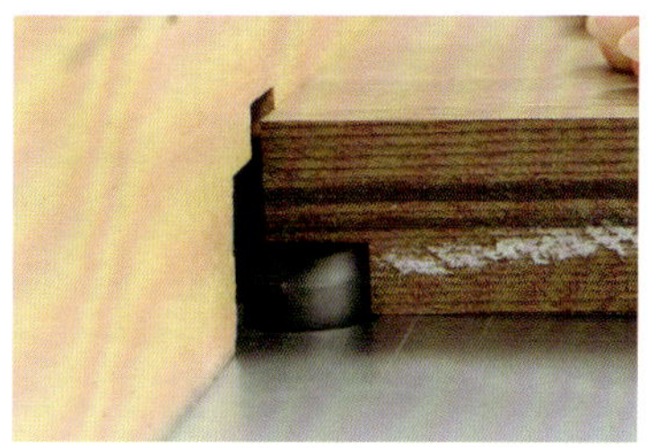

DOVETAIL BITS CAN'T CUT IN SEVERAL PASSES

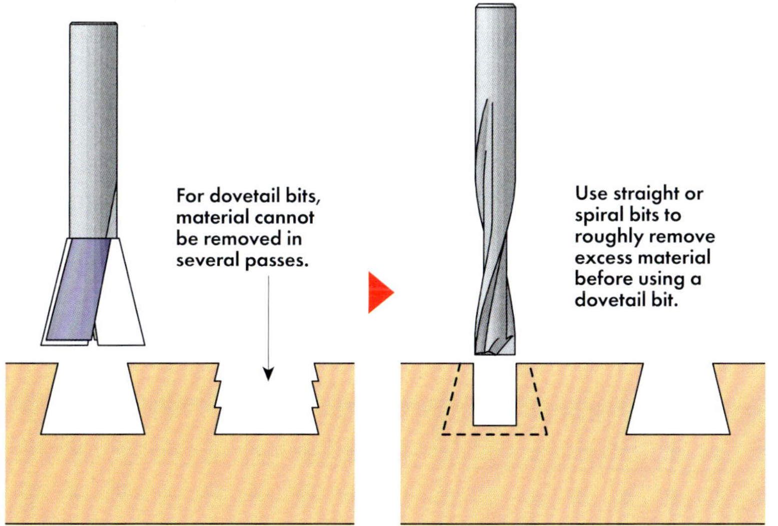

Constructing a Trimmer Table

The router table being constructed here uses the MIRAI Trimmer Table Set for its tabletop, but similar router tabletops should be available for you to purchase. The tabletop is made of high-precision aluminium, ensuring flatness.

It features two bit holes of different sizes, allowing for differentiation based on bit diameter. It doesn't use an insert ring system, so feeding material is very smooth. Changing between the large and small bit holes involves a structure that allows the router to slide internally.

The height of the stand has not been given here, as it varies for each router. The method for deciding on stand height is based on the length of the router with a straight bit attached. This ensures that the bit can be lowered to the surface of the tabletop.

The interior is lined with soundproofing sheets to reduce noise. This isn't mandatory of course, so judge for yourself whether or not it is necessary based on your working environment.

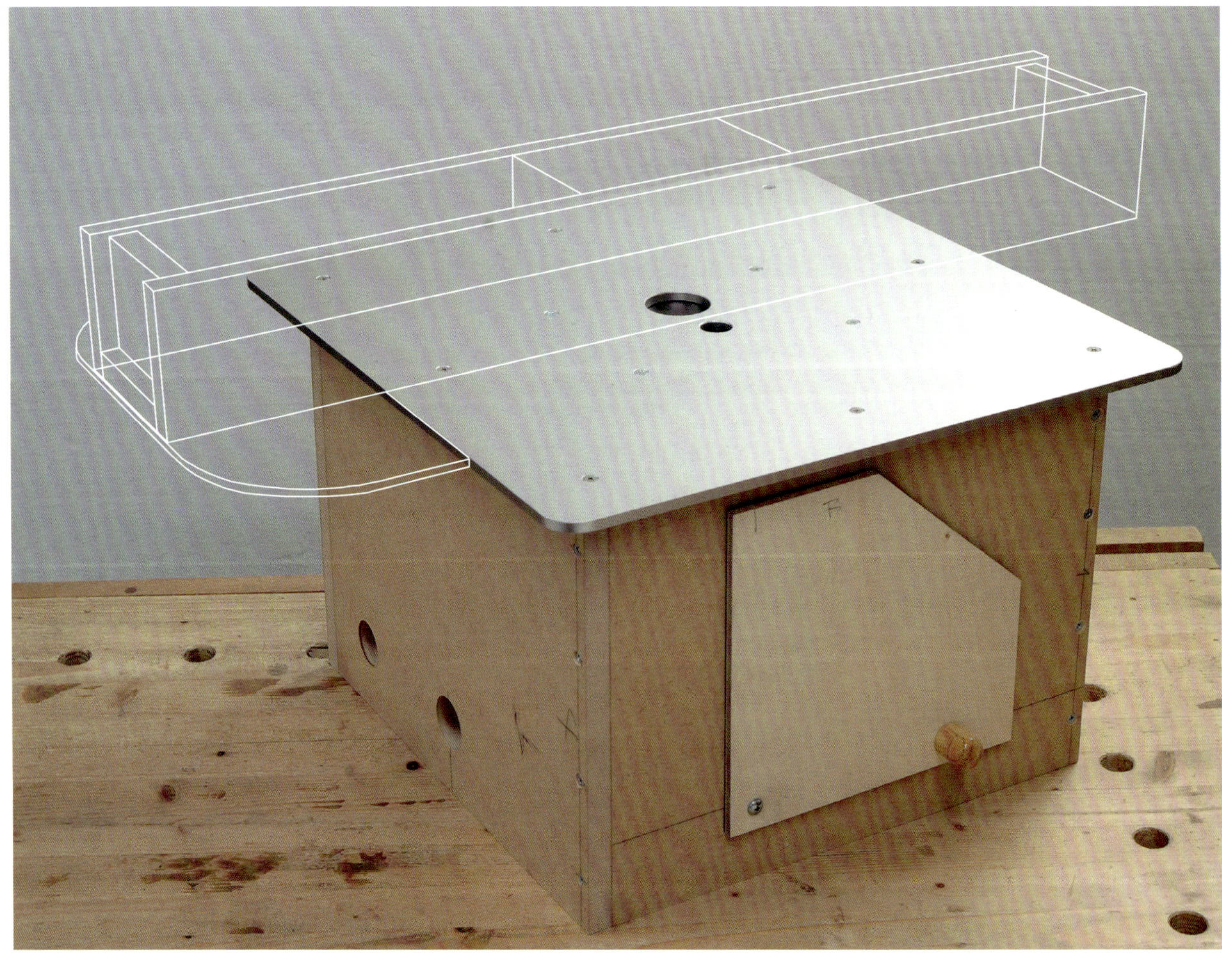

◉ SWITCHING BIT OPENINGS BY MOVING THE SLIDING BASE

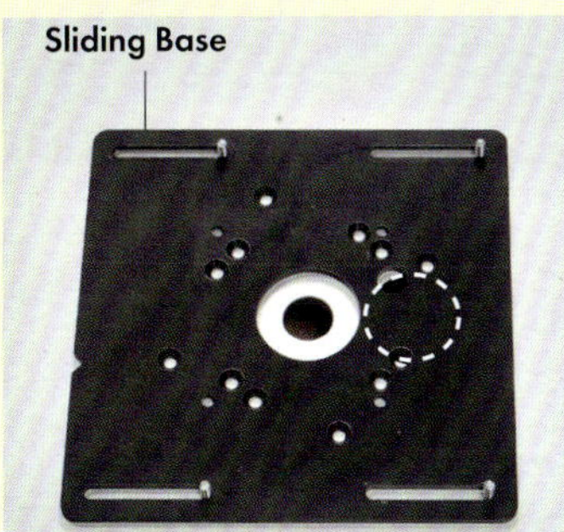

This is a photo of the tabletop's underside. It features two bit openings (large and small). Here, the sliding base is positioned over the small bit hole, while the large bit hole is concealed (indicated by the dashed line).

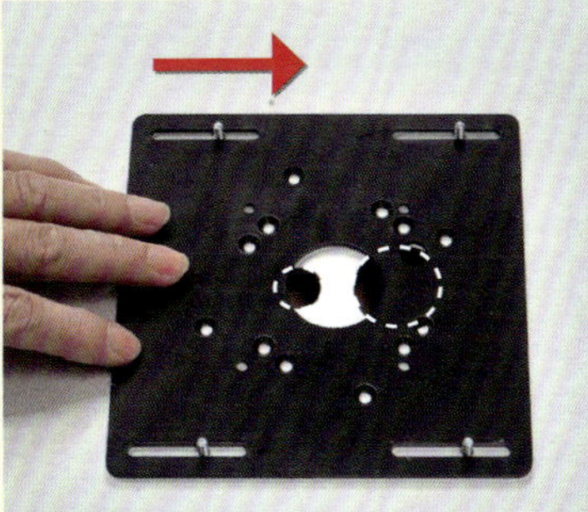

Move the sliding base.

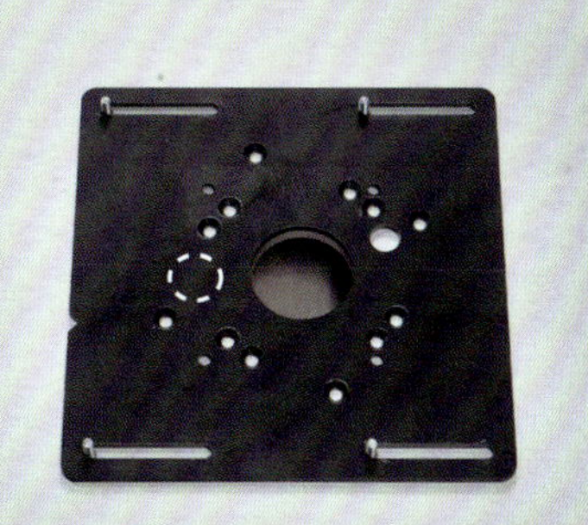

The sliding base has been moved to cover the large bit opening. The small bit hole is now concealed (indicated by the dashed line).

During actual use, the trimmer is attached to the sliding base.

Fence/Stand Wood Cutting Dimensional Diagram

Units: mm/t: thickness

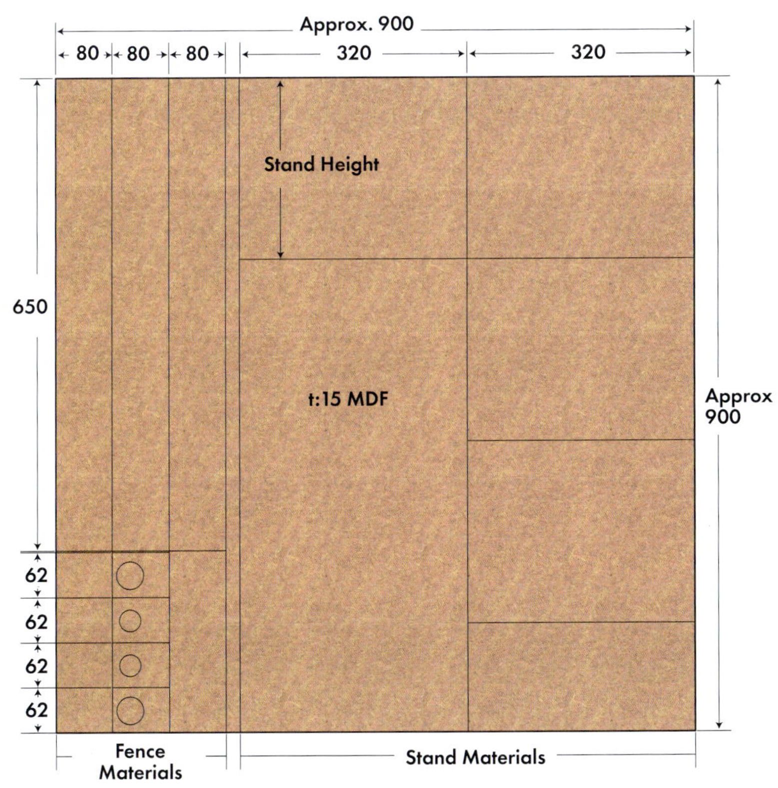

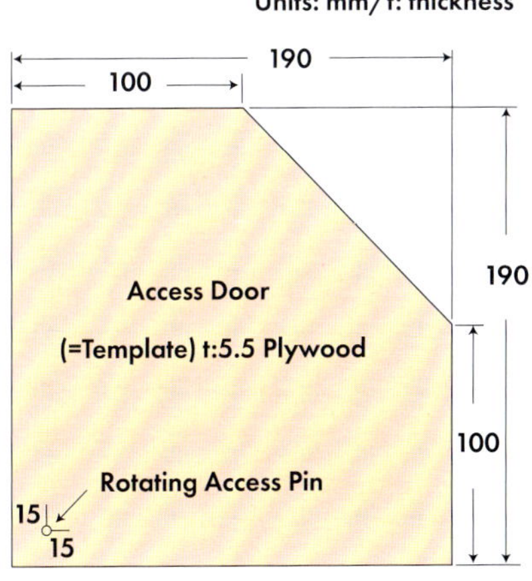

Two sheets of plywood are necessary.

Stand Construction

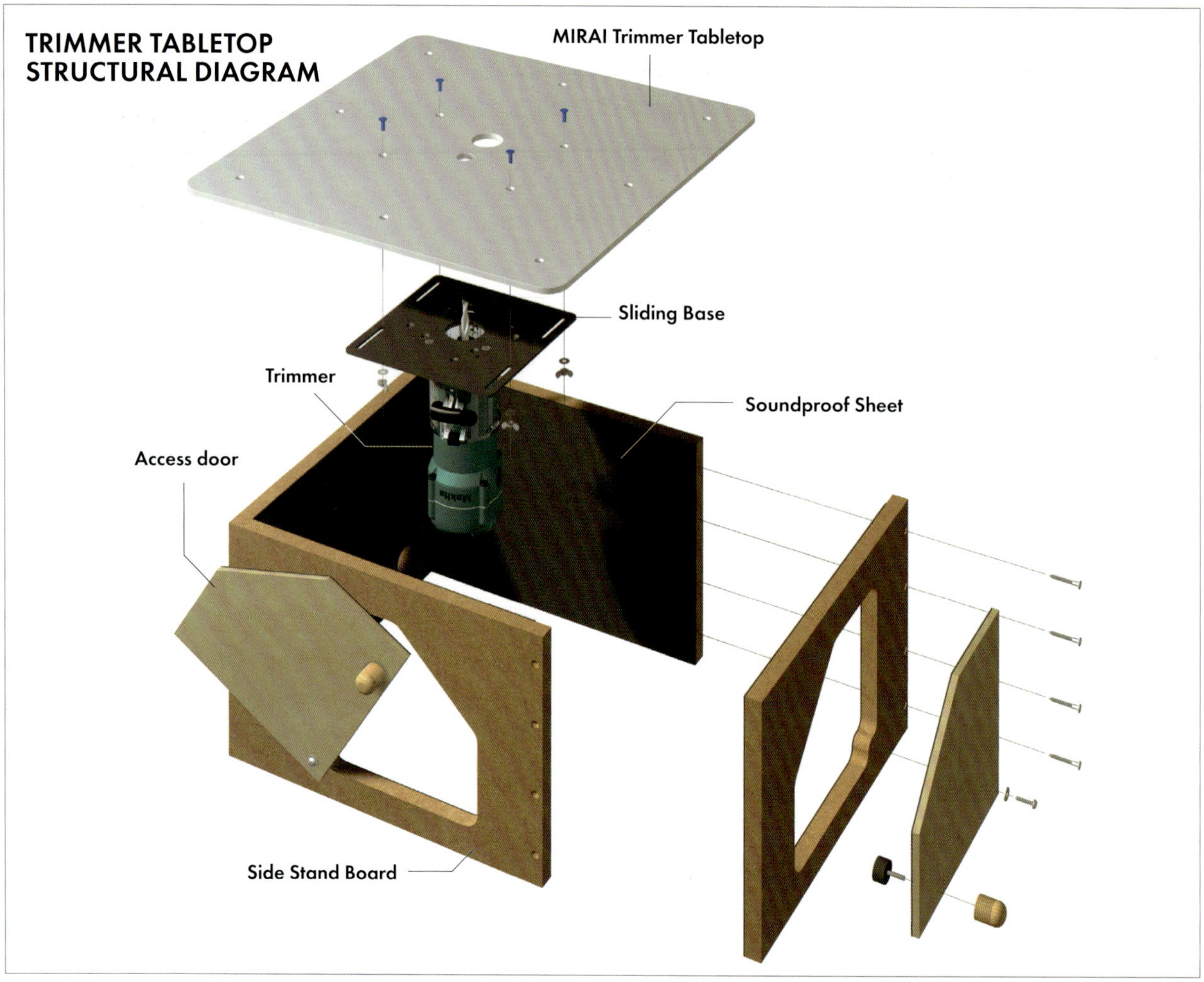

TRIMMER TABLETOP STRUCTURAL DIAGRAM

MIRAI Trimmer Tabletop
Sliding Base
Trimmer
Soundproof Sheet
Access door
Side Stand Board

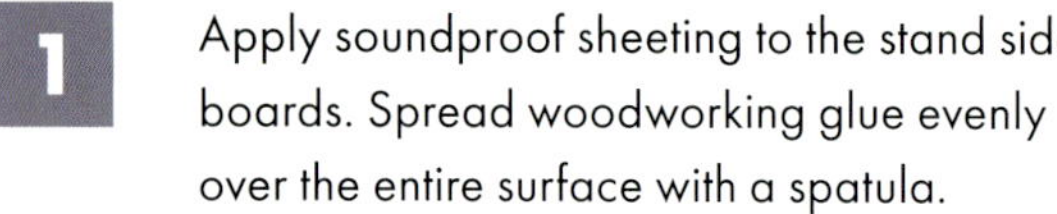

1 Apply soundproof sheeting to the stand side boards. Spread woodworking glue evenly over the entire surface with a spatula.

2 The stand board uses butt joints, as shown. Be sure to leave enough space when positioning each piece of soundproof sheeting (notice the space near my knuckle) to account for board thickness plus soundproof sheeting.

3

Apply masking tape to areas where soundproofing will not be attached. This will make it easier to clean up any excess glue that may ooze out.

4

Stack everything and clamp as tightly as possible using as many clamps as possible. Be sure not to place clamps on the areas without a soundproofing sheet. Leave to dry overnight.

5

Once the glue has set, trim off any excess soundproofing sheet.

ACCESS DOOR OPENING

6

Place the access door template on the stand board to determine its position. Insert a pushpin where the rotating axis pin will go. Make sure that the door doesn't hit the top board when it rotates.

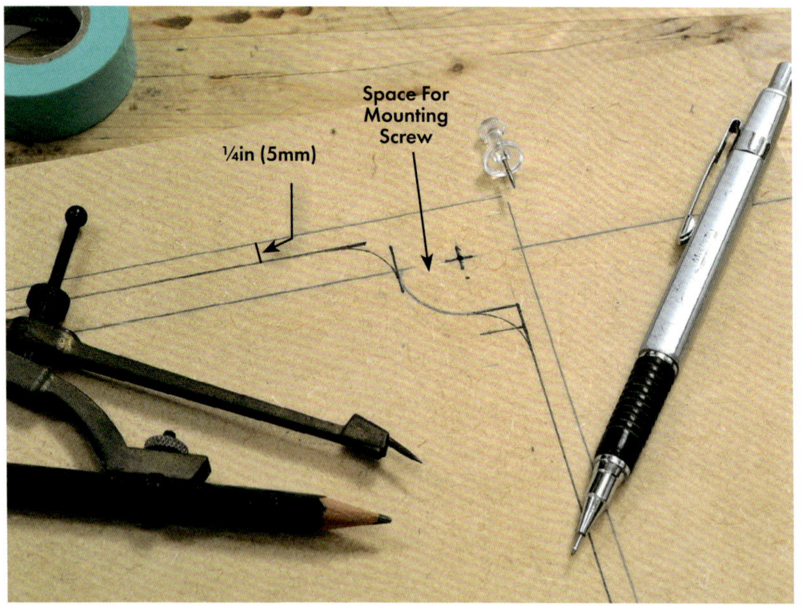

7

Once the door positions are set, trace doors onto stand boards. Draw lines ¼in (5mm) in to set the actual opening size. Make sure to leave space for mounting screws, especially around the rotating axis pin, as shown.

8

Temporarily assemble the structure with access doors on the front and right sides.

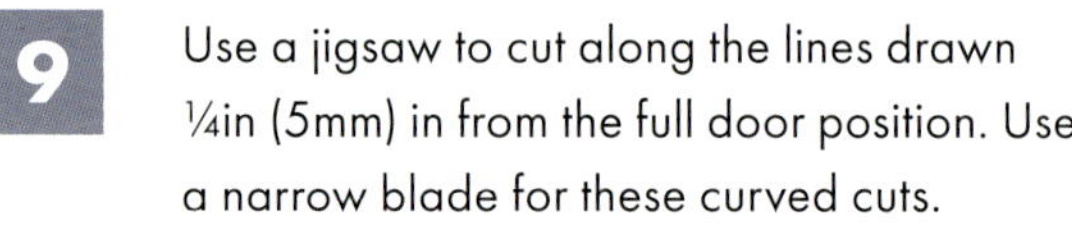

9

Use a jigsaw to cut along the lines drawn ¼in (5mm) in from the full door position. Use a narrow blade for these curved cuts.

10

Cut through the soundproofing sheet as well. The second piece should be cut in the same way.

11

The piece is cut cleanly. There are soundproofing sheet remnants on the cut edges, making them appear black.

CONSTRUCTING AN ACCESS DOOR

12 Plywood just over ¼in (5.5mm) is cut out, and soundproofing is adhered to it.

13 Clamp the two pieces together. After leaving them overnight to set, trim off any excess soundproofing.

INSTALLING THE KNOB

14 Attach the knob and some door stop rubber, on the back, to access the door. Determine the installation positions from both inside and outside. The door stop rubber will fix the door's position when opening and closing.

15 Once the positioning looks good, attach the knobs.

16 Temporarily assemble the box and check the door positions while opening and closing.

REINFORCEMENT HOLE

The door's pivot uses wood screws. However, over years of use, the hole may become enlarged and become sloppy. To prolong its lifespan, I cut a brass pipe and embed it in the hole. This can be done later if necessary.

STAND BOARD ASSEMBLY

18

Drill holes for the dust collection hose and trimmer cord. Draw a centreline on the wood's edge where the trimmer tabletop will sit. This line should be visible through tabletop screw holes. Additionally, mark pilot hole positions. These will be used to screw the stand boards together with butt joints.

19

Dry fit the two boards. While holding them firmly in place with corner clamps, drill pilot holes, and then screw them together. Since MDF edges are prone to splitting, be sure to drill pilot holes!

20

The two L-shaped assembled stand boards can now be joined together to form the box. In this build, no adhesive was used, but using glue would make the box even more robust.

TABLETOP INSTALLATION

21 Remove the aluminium tabletop from the trimmer tabletop set.

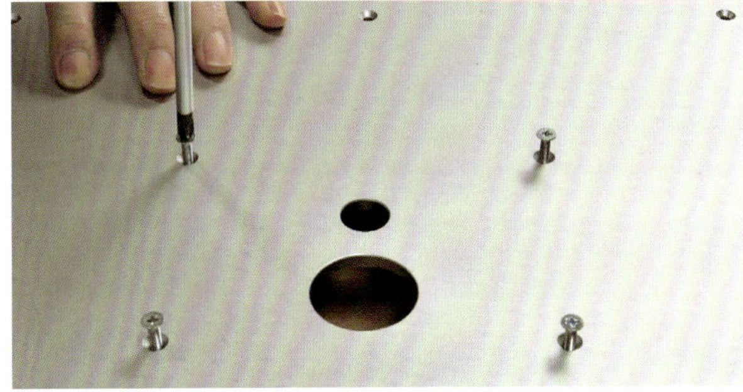

22 Drill the provided screws through four central holes.

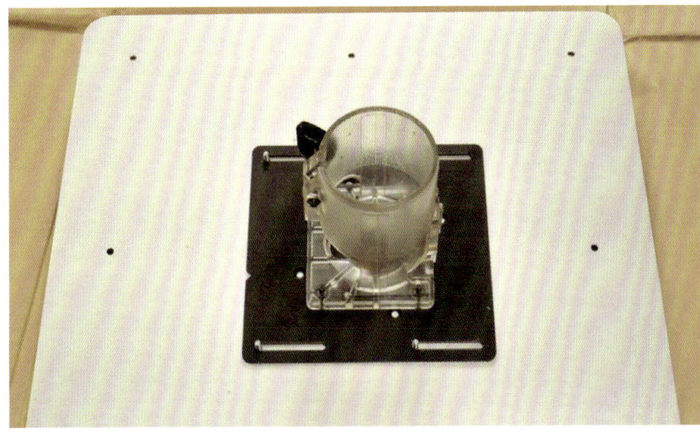

23 Flip the tabletop over and place the sliding base on four screws. Attach the trimmer housing on the sliding base in advance.

24 The instruction manual advises that, in the final step, the sliding base should be surrounded by scrap wood pieces secured with double-sided tape. This prevents the sliding base from shifting too far side to side and ensures precise work when making box joints.

25 Place the tabletop on stand panels and secure with wood screws. Attach the access door to complete assembly.

Constructing a Fence

Instructions for the DIY fence in the product manual utilize toggle clamps. However, we modified the instructions to use standard clamps. Additionally, while the manual specifies a width of 2⅜in (60mm) for the bottom plate, I made it 3⅛in (80mm) for this project.

The fence will be made from 9/16in (15mm) MDF, similar to the stand. There are fence boards at both the front and back, and one side of the fence features a cut-out for a dust collection port. This allows for dust collection when processing material edges. On the other hand, for groove processing, the fence without any cut-out will be used. There is no cut-out while feeding material on this side of the fence so there isn't any snagging. This allows for smoother operation. As there is no cut-out, dust collection will be performed through the stand.

The dust collection hose is routed off to one side instead of directly along the centreline of the fence. This allows for additional clamping space (see photo 9 on page 95). It is secured to the fence top using C-clamps or F-clamps. As you can see, the dust collection hose can be attached from either side.

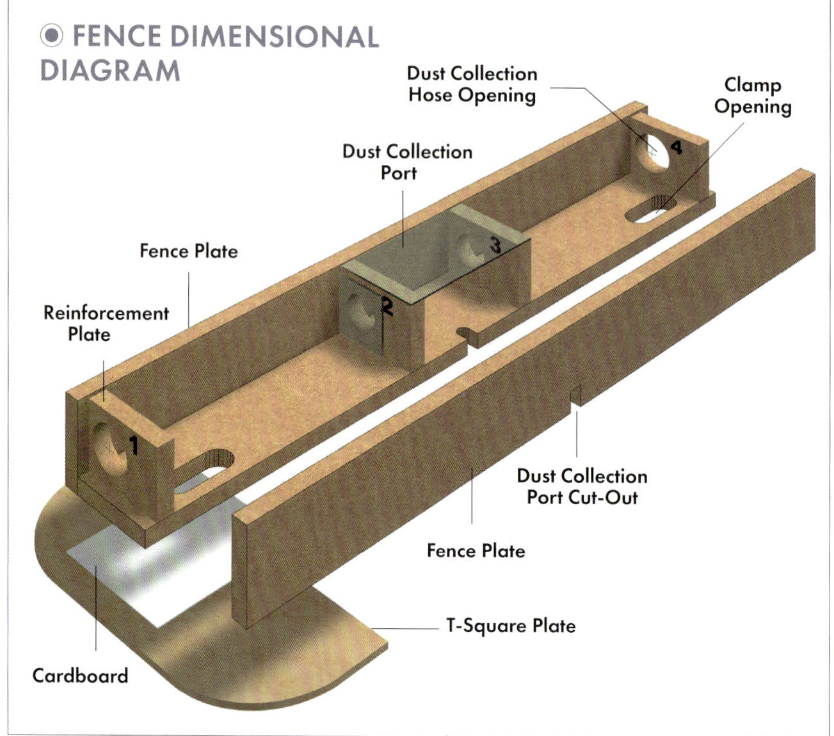

REINFORCEMENT PLATE AND DUST COLLECTION HOSE OPENING

1

Before cutting out your reinforcement plates, drill holes for the dust collection hose. Be sure to label them with numbers 1 through 4. When gluing, arrange them in that order. There are two sizes of hole: Holes for the hose (1 and 4) should be approximately ¼in (5mm) larger than the hose fitting diameter, while the holes for inserting the fitting (2 and 3) should be snug. After drilling the four holes, cut the plates into four pieces.

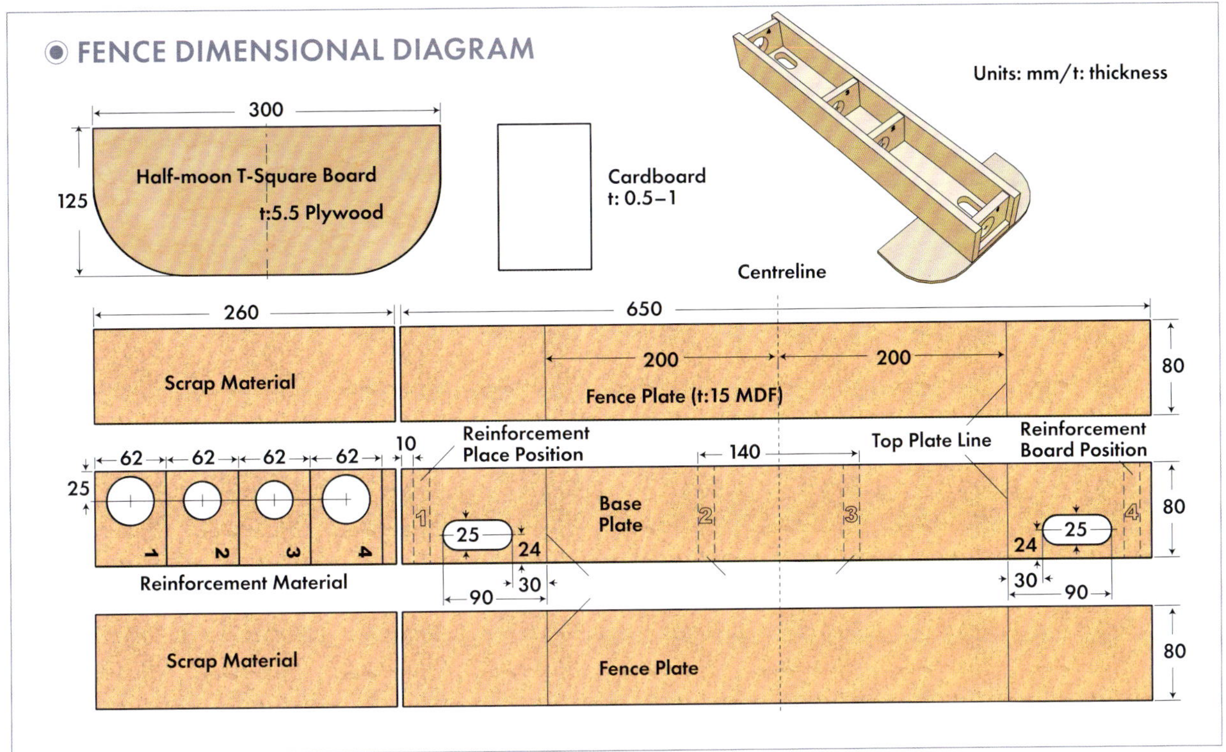

2

The photo shows the cutting of reinforcement plates to the same size with a hand saw. The author has published a book on woodworking that doesn't use electric tools. It emphasizes a quiet and precise approach to woodworking.

3

Holes are drilled and reinforcement boards are cut as shown. A Forstner bit was used to drill out vacuum hose holes. However, it's also possible to use a hole saw, as seen in the top right corner.

BASE AND LONG SLOT PROCESSING

4 Next, drill elongated clamp holes in the fence base plate. Here, a 1in (25.4mm) Forstner bit is used to drill continuous holes that are connected into an elongated hole. This hole is just for a clamp, so it's acceptable if the straight edges aren't perfectly finished. First, drill holes at both ends and then connect them.

5 The fence length is 25½in (650mm). Draw a centreline. Then, draw lines 8in (200mm) to the left and right. These will indicate the trimmer top (labelled 'Top Plate Line' in the Fence Dimensional Diagram). Mark positions for reinforcement plates 1 through 4.

GLUING

6 Prepare everything, including clamps and other materials, before starting the gluing process.

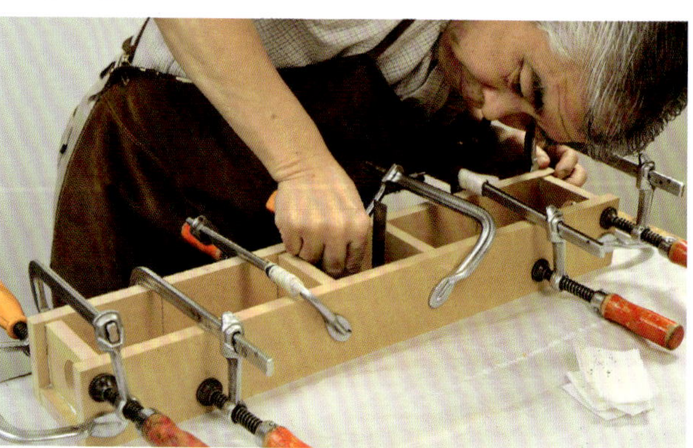

7 Assemble the entire structure and clamp. Ensure the bottom panel doesn't lift by pressing it down firmly. Check that the reinforcement boards remain level.

8 Place an adjustable T-square against the inside of the fence to confirm that it is square. Squareness depends on right-angle accuracy when cutting the reinforcing board.

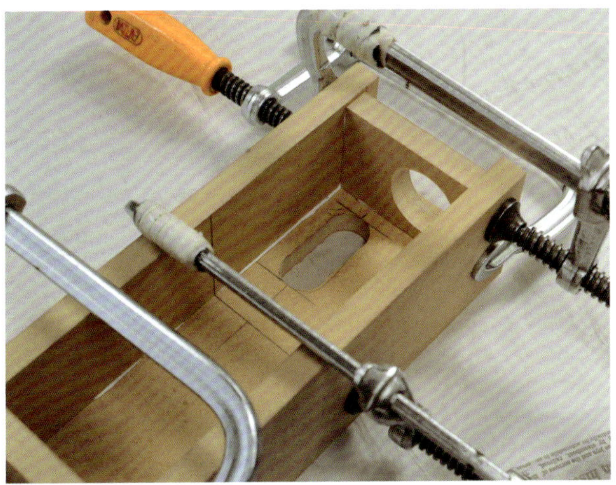

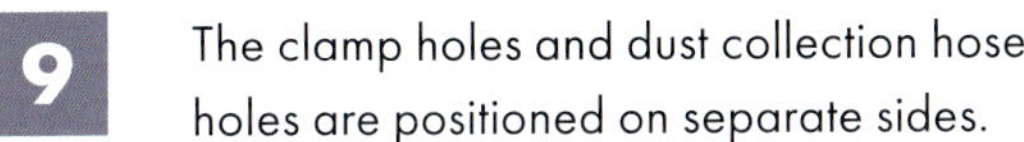

9 The clamp holes and dust collection hose holes are positioned on separate sides.

10 Clamps are placed adjacent to reinforcement plates.

CONSTRUCTING A T-SQUARE

11

The fence bottom plate and semicircular T-square board are glued and screwed together. However, they are glued with a piece of thick paper or plastic between them. This prevents the T-square board from dropping below the tabletop, thereby preventing material from snagging (see step12).

12 The photo shows the difference between the aluminium top and T-square board. The T-square surface is lowered by the cardboard. The cardboard is 0.5mm thick, but 1mm would work.

13 Apply woodworking glue to the cardboard and attach it to the fence bottom. Cut it down to 1mm smaller than the fence width.

14

Glue the T-square board to the cardboard and screw in place. The board has a centreline, fence width lines, and clamp hole positions drawn on it. This is for determining the wood screw positions.

15

Use the trimmer again to make a long hole. Use a flush trim bit with bearings for this process. First, drill a bit hole, and then proceed to cut so that this long hole is aligned with the existing long hole on the other end.

16

Both clamp holes, of the same size, have been drilled. Remain cognizant of the wood screw positions used to secure the T-square board.

DUST COLLECTION PORT

17

Cut a transparent PVC sheet (3⅛×5½in/80×140mm, thickness: 5/64–⅛in/2–3mm) using an acrylic cutter.

18

Attach to the reinforcement plate with double-sided tape or screws. Use acrylic to seal one of the hose holes (hole 2 of the reinforcement plate).

CREATING A DUST COLLECTION PORT

19

To collect shavings, it is necessary to cut a port in the fence. This can be done when using the fence for the first time.

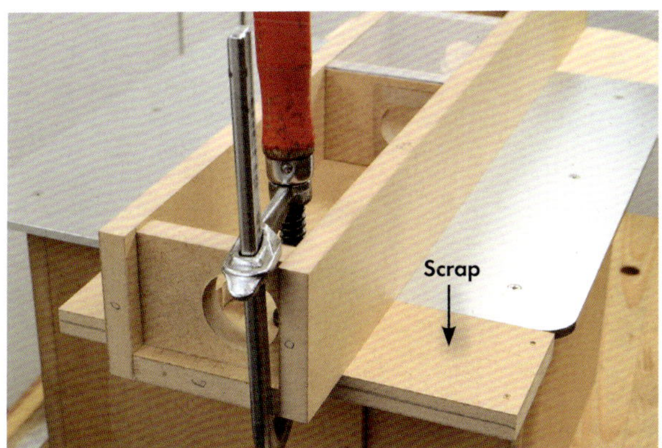

20

Clamp a piece of scrap to the fence end so that the entire fence can slide back and forth easily.

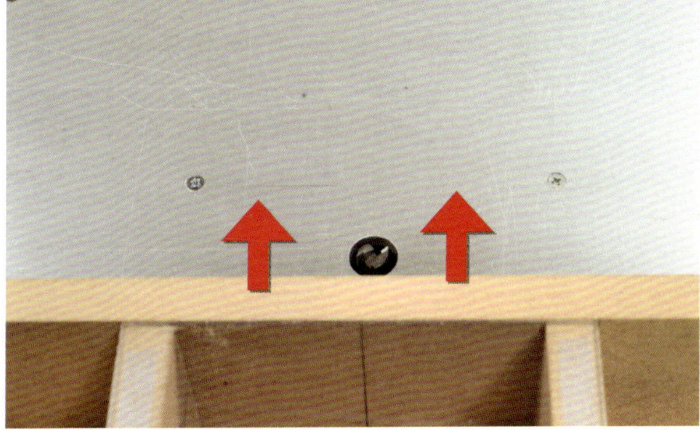

21

Attach a commonly used bit to your trimmer and slide the fence to create a notch. Raise the bit height until the notch appears inside the dust collection port.

22

The port is now opened, thus allowing for dust collection. As you raise the bit higher and higher, the notch will grow. These notches will expand with each project you create. This photo has the transparent PVC sheet removed for demonstration purposes.

4

Chapter 4

Trimmer Projects

Simple Mitre Joint Box 2

Inset Inro Box

On page 58, I introduced a box with a large mitre joint. It was made with a method similar to folding cardboard. This time, I will make a box with a lid using the same method. First, I will create a sealed box. Then, I will cut it into both a lid and a body.

We will create a protrusion called an inro to align the lid with its original position. The methods for making an inset inro and an attached inro are different. First, we will only go over the inset inro. The method for making the attached inro will be introduced from page 108. Both will utilize the trimmer table. In the explanation on page 58, the central portion of the MDF board served as the bottom of the box. This time it will serve as the top. The bottom will be made of 5⁄32in (4mm) plywood. The trimmer bits used include a V-groove bit and a 1⁄8in (3.2mm) up-cut spiral bit.

STRUCTURAL DIAGRAM • UNFOLDED DIAGRAM

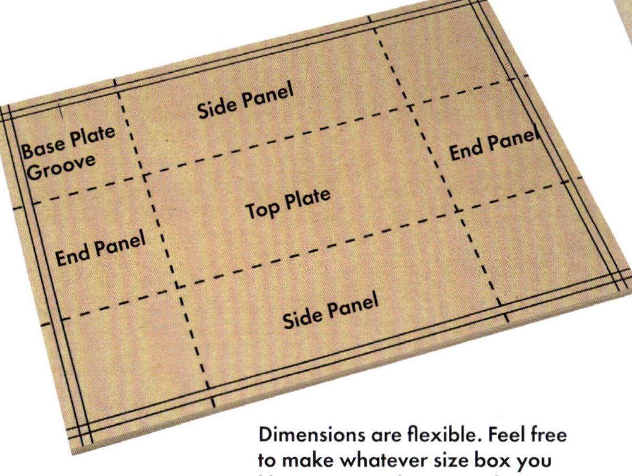

Dimensions are flexible. Feel free to make whatever size box you like. However, be sure to keep the height of the side panels and end panels the same, as this is absolutely necessary.

ATTACHED INRO AND INSET INRO DIAGRAMS

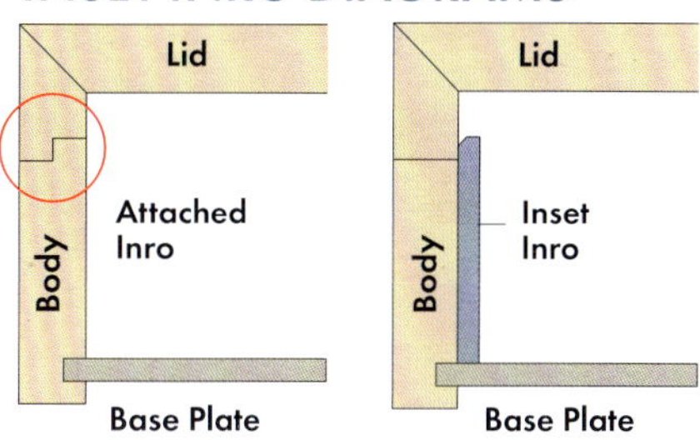

PRODUCTION STEPS

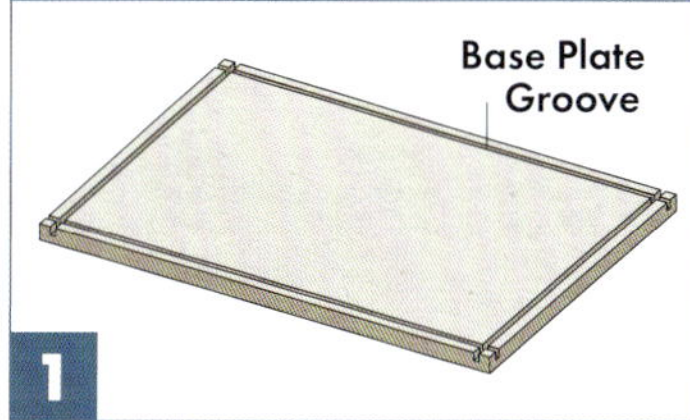

Cut the groove for the Base Plate.

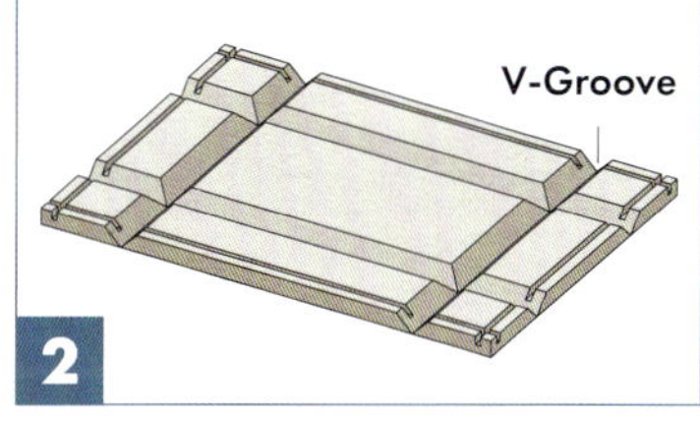

Use a V-groove bit to cut deep grooves at the fold locations.

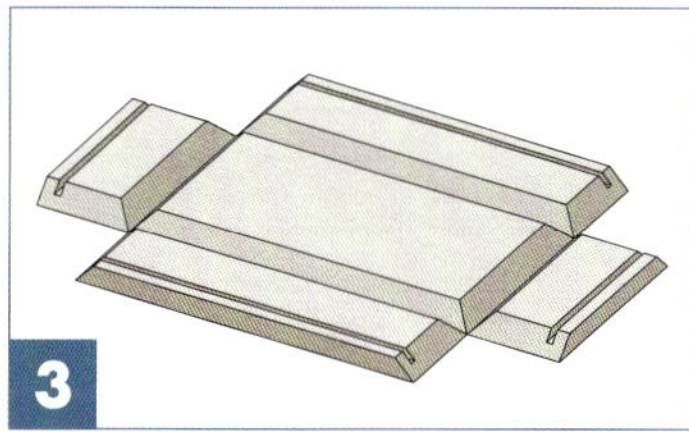

Cut out the four corners with a box cutter.

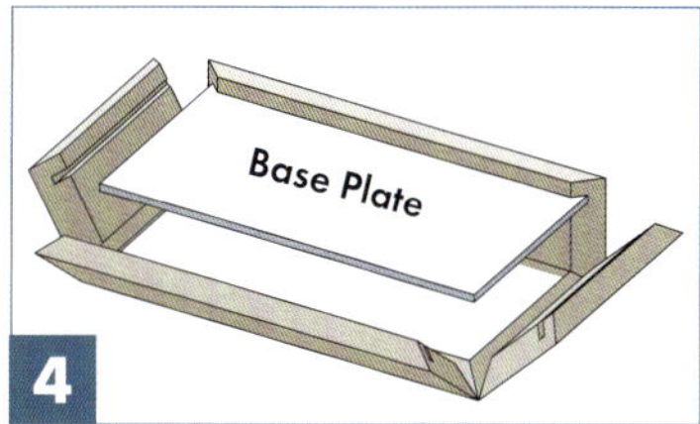

Fit base plate while gluing the mitre joints.

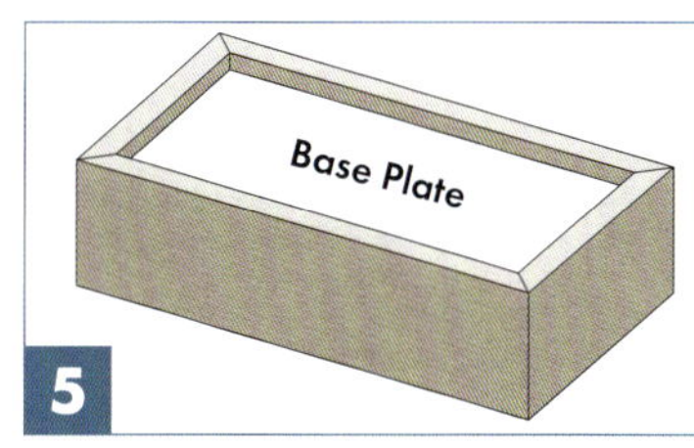

A sealed box is formed.

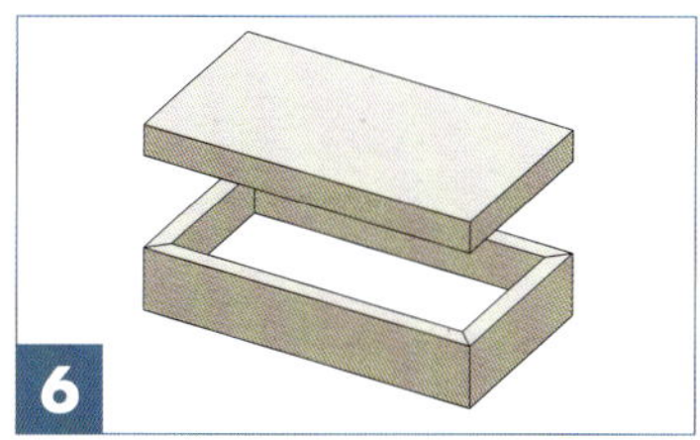

Cut the lid and body apart.

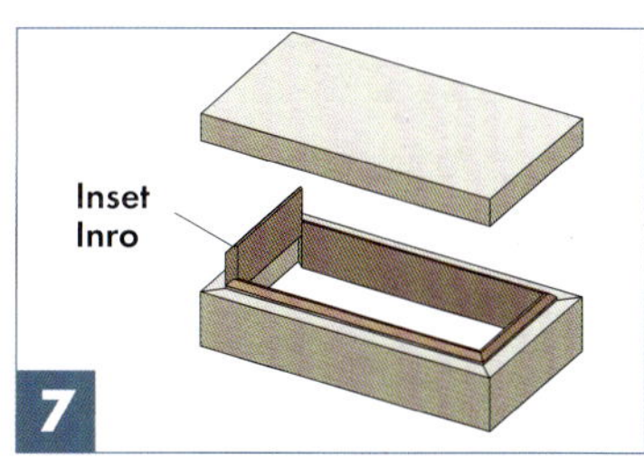

Press in the inset inro and it's done!

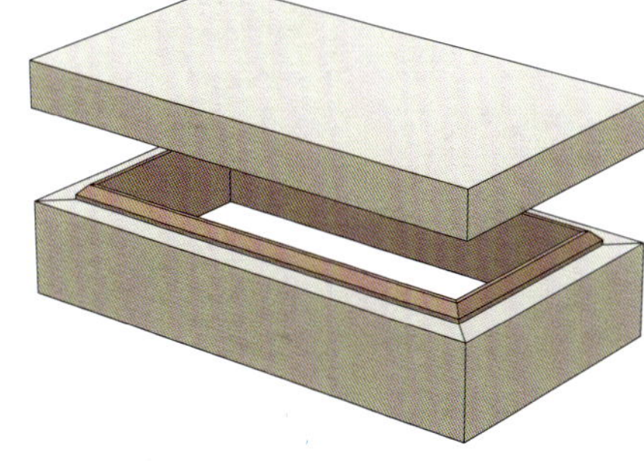

CUTTING THE BASE PLATE GROOVES

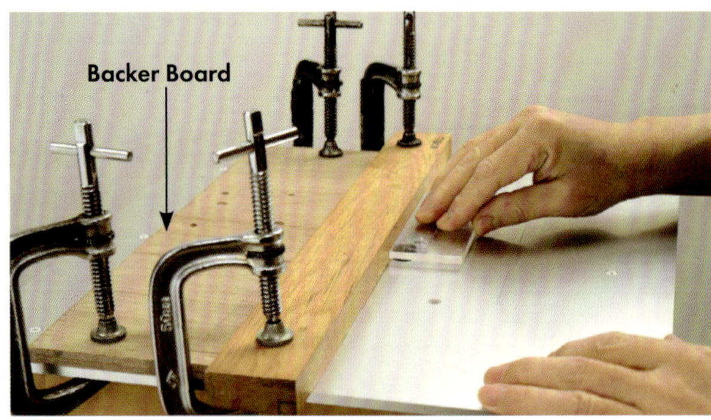

1 Use a spiral bit to cut a groove ¼in (5mm) from the bottom edge of the box. Place a ¼in (5mm)-thick acrylic spacer between the fence and bit to set groove position.

2 Place a backer board behind the fence and secure with a clamp. This step is essential for groove widening.

3 Next is depth setting. Here, we'll set the depth to ⅛in (3mm). Place a ⅛in (3mm) acrylic sheet on the router table as a spacer. Then, place a wider, thicker acrylic sheet on top as a stopper. Push the spiral bit up from below and set to ⅛in (3mm) by pressing the bit against the stopper.

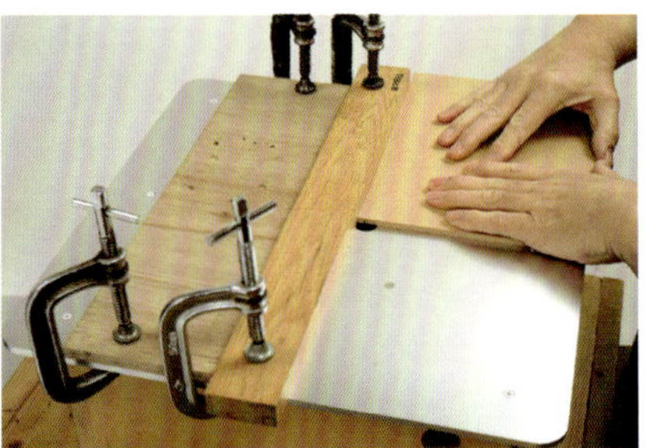

4 Cut grooves into four sides of the material.

5 First pass is complete. The groove width is currently set to ⅛in (3mm), but we are using 5/32in (4mm) plywood for the base plate, so the groove needs to be widened by about 3/64in (1mm). Loosen the clamp on the backer board, insert a 3/64in (1mm) spacer between the clamp and fence, and then re-secure the backer board.

Next, remove the clamp and take out the spacer. Then, position the fence directly against the backer board and secure. This pushes the fence back by 3/64in (1mm). Note: When widening grooves, always move fence the backward. Moving it forward is dangerous (see page 82).

7

Process all four grooves again to widen. The groove cutting is now complete.

V-GROOVE PROCESSING

8

Process the V-grooves in a similar fashion. It's common not to hit the desired depth in one pass, so be sure to make several passes. During processing, it is absolutely necessary to use a push pad or scrap material to prevent the workpiece from lifting and protect your hand. As you approach maximum depth, apply masking tape to the underside.

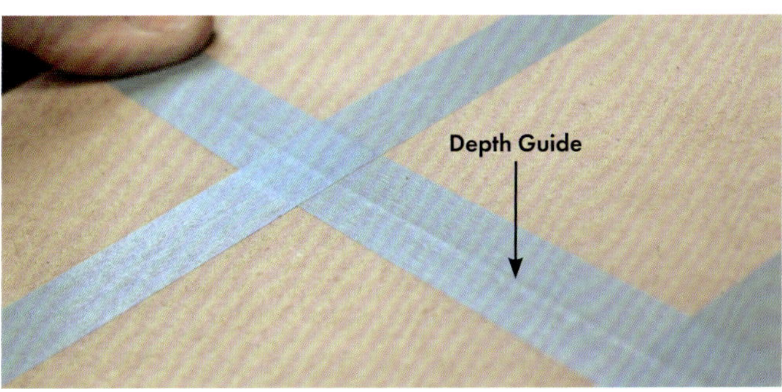

V-grooves are processed right up to the material thickness. A good depth guideline is when the deepest part of the V-groove appears as a single raised thin line on the masking tape. If you think it would be better to increase the depth a little more for folding, feel free to raise the bit height by a small amount.

All grooves are processed.

10 Next, cut four corners out with a box cutter and temporarily fold to assemble. Check the mitre joint positions to make sure all the sections and grooves align.

BASE PLATE

11 Determine the base plate size. Add groove depths on both sides, and subtract 3/64in (1mm) from the total. By cutting it 3/64in (1mm) smaller, you can prevent the base plate from interfering and causing gaps in the mitre joint sections.

12 Insert the base plate – cut from 5/32in (4mm) plywood – and check for any issues.

GLUING

13 Apply adhesive. If you apply too much, it will overflow inside the box! So, use an appropriate amount. Here, the base plate groove is not coated.

14 Insert the base plate while closing the final side panel.

15 Use thick rubber bands to firmly tighten. Check for right angles with a square. If it's out of square, clamp diagonally to make adjustments (see page 61).

16 The sealed box is complete.

17 Once the glue has dried, correct any slight unevenness on the box bottom with sandpaper. Place the sandpaper on a flat surface and sand down the edges. After this, separate the lid and body.

SEPARATING THE LID AND BODY

18 We'll separate the pieces using a spiral bit. Tip: Set the bit height slightly above side panel thickness to completely cut through them. For the end panels, adjust the bit height and work in several passes. Leave about 0.5mm uncut so that you can finish cut later with a utility knife. If you try to cut through all four sides at once, the lid and body may start moving, making it difficult to achieve a clean cut.

19 Cut the end panels apart with a box cutter.

20 Clean cuts are made. Next, we will process the inset inro.

MITRE JOINT INSET INRO

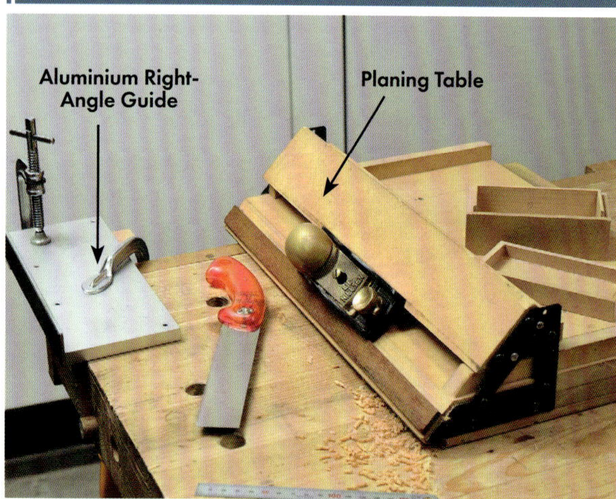

21

The inset inro is a delicate and detailed item, so hand tools are used instead of power tools. The aluminium right-angle guide on the left is used for cutting right and 45-degree angles (see page 233). Here, we will utilize its right-angle cutting function. On the right is the planing table. It allows for planing at a 45-degree angle. The black triangular component is a commercially available item known as a 'planing table bracket'. Wooden parts are custom-made to fit your specific needs. The plane shown in the photo is Western style.

◉ INSET INRO CORNER PROCESSING

The inset inro protrusion serves to align the lid and body of the box. Here, we will create the 'mitre joint inset inro' shown on the left. This method requires a bit more effort, so you could also opt for the 'butt joint inset inro'. When using a butt joint, the inset inro is glued to the inside of the box. By the way, the author typically skips the glue when creating a mitre joint inset inro.

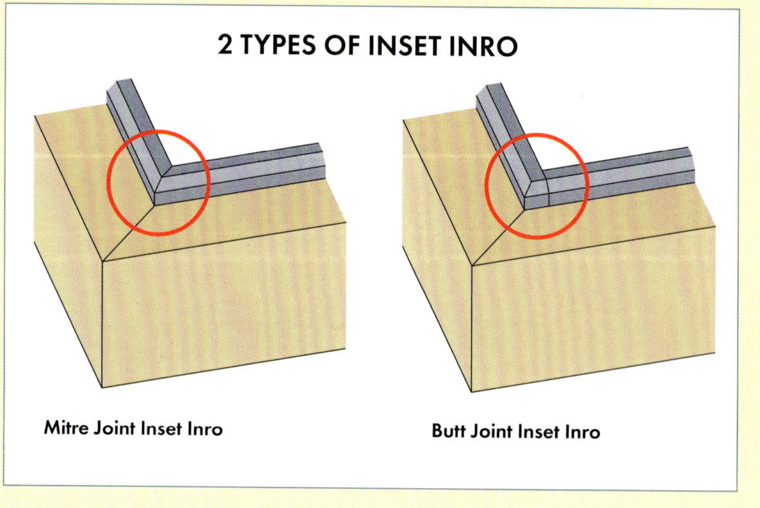

CUTTING INSET INRO

22 Cut the inset inro material to a width that extends about ¼in (5mm) above the top edge of the body. Here, I'm using ⅛in (3mm)-thick wood. You can just make light cuts on each side using a marking gauge. Bend along these cuts to achieve the desired dimensions.

23 Use the guide squarely to cut the material to length. The saw blade sticks to the magnetic sheet on the guide surface, ensuring a straight cut with a square cross-section.

45 DEGREE PROCESSING

24 Press the material against the planing table and plane a 45-degree angle. Gradually shave both ends while adjusting to fit box.

25 If you firmly insert the last board, the inro should fit snugly.

26 Adjust the fit of the lid using a plane or sandpaper, and it's complete.

Simple Mitre Joint Box 3

Attached Inro Box

We learned about inset inro on page 100. Here, we will create a box with an attached inro. An attached inro is formed as a single unit from the box material, including the inro part. Please refer to page 101, 'Attached Inro and Inset Inro'.

For the inset inro, we only processed two grooves: one for fitting the base plate into the material and another for the V-shaped folds. However, for the attached inro, we'll need to process an additional groove where the lid and body will be separated. In total, we will process three grooves. After that, we will glue up the sealed box.

Up to step 9 on page 103 everything is the same. We will explain operations from there.

TRIPLE GROOVES

◉ PROCESSING STEPS FOR ATTACHED INRO: OUTER BOX PARTS

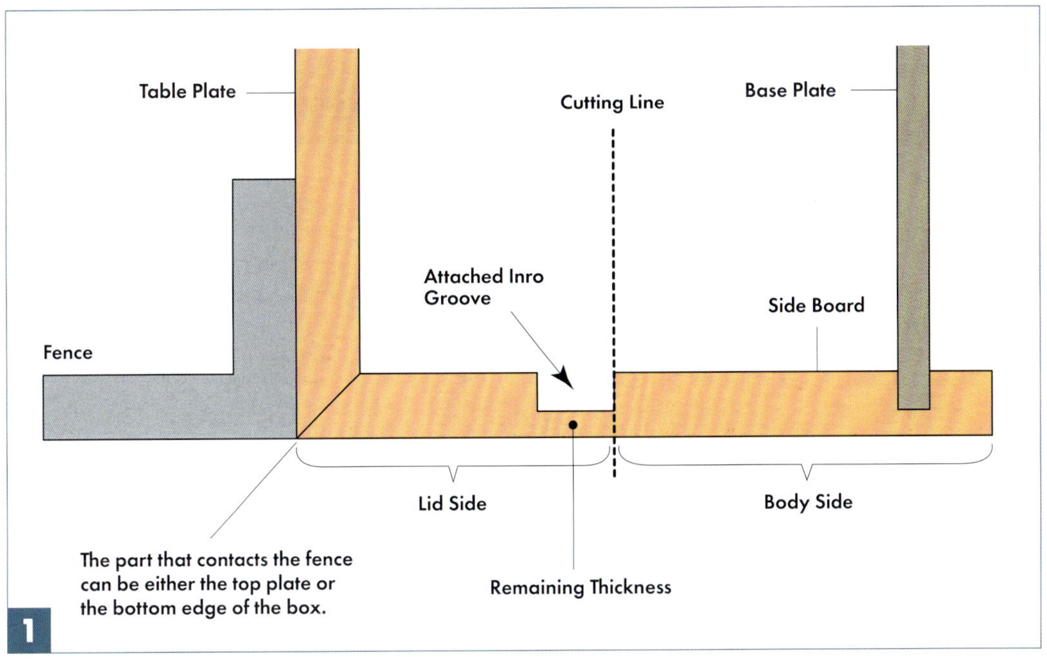

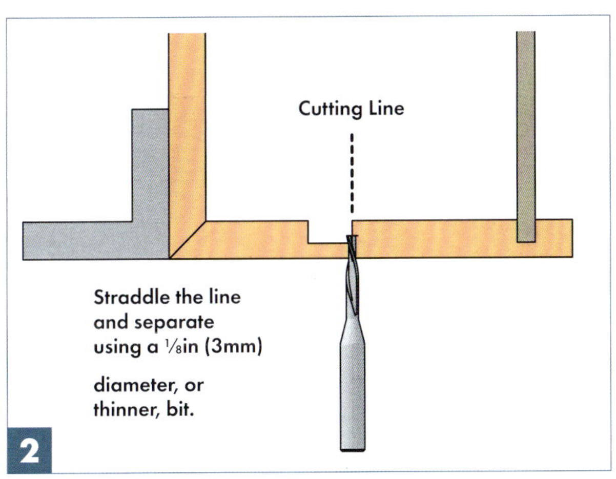

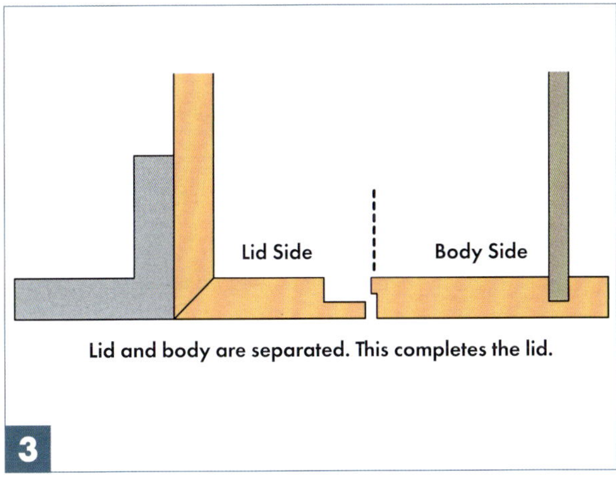

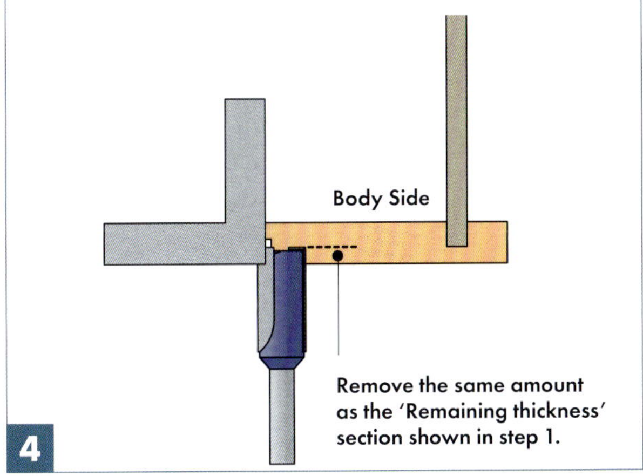

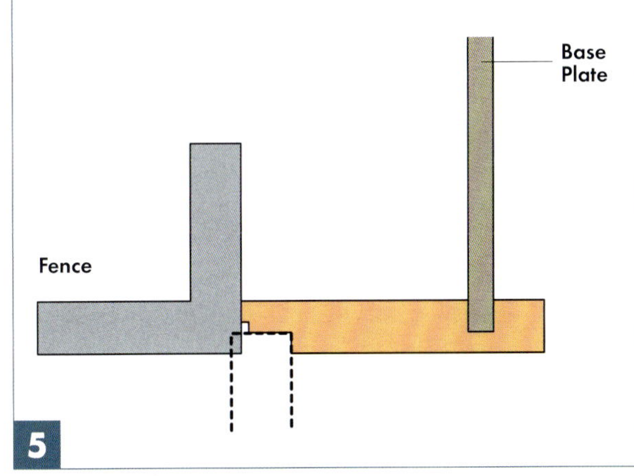

Next, we will process the attached inro grooves, after base plate grooving of course. The router bit used will be a ¼in (6.35mm) down-cut spiral bit. The groove depth is about half the material thickness. The pencil line visible to the right of the groove indicates the V-groove position.

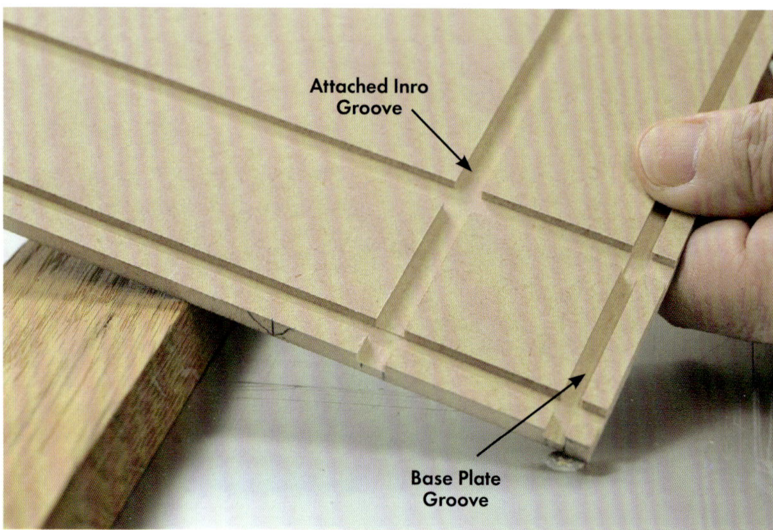

2

After processing grooves, flip the material over. The inner groove seen here is for the attached inro.

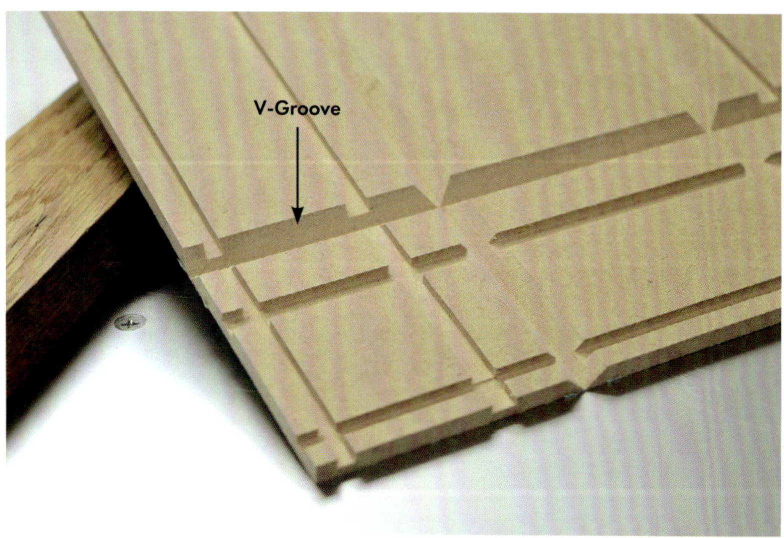

3

All groove processing, including the V-grooves, is complete.

4

After cutting out four corners with a box cutter, apply masking tape as shown. When assembling, glue will be applied to all 45-degree surfaces. Careful as this may cause overflow. The overflow affects the inside of the sealed box, so it is difficult to treat. Apply masking tape to make it easier to remove the glue once it has hardened.

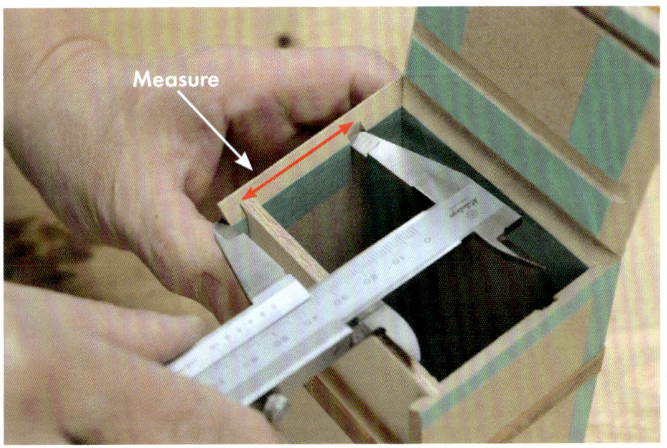

5 During dry assembly, measure the distance from the attached inro groove to the bottom edge. Make a note of it. This distance will be needed when cutting apart.

6 Next, measure the remaining part thickness. Measure from the attached inro groove. This thickness will determine the amount to be cut off the outside edge of the box.

7 Apply glue and assemble.

8 Check for square corners.

9

Once the glue has set, smooth the box bottom with sandpaper. This area will be against the fence when cutting apart the lid and body, so any unevenness will cause inaccurate machining.

CUTTING APART THE ATTACHED INRO

10

Mark the distance from the inside groove of the attached inro to the bottom edge of the box – as in step 5. Set the trimmer bit to cut along that line. Use a ⅛in (3.2mm) down-cut spiral bit.

11

Press the bottom edge of the box against the fence and cut. Don't push the lid side against the fence. Only press the body against the fence to cut. Unlike method 18 on page 105, which leaves a thin layer on the end panel to be cut later with a knife, this method involves cutting all the way through.

12

This photo was taken immediately after cutting. As you can see, there are burrs. They will need to be cleaned up with sandpaper or a knife. The lid is on the right side and you can see that the inro has already been completed.

Remove the masking tape. Excess adhesive has solidified, but thanks to the tape, it can be removed relatively easily. On the right we see the body. It has a slight step where it was cut apart, but this will be sanded down later.

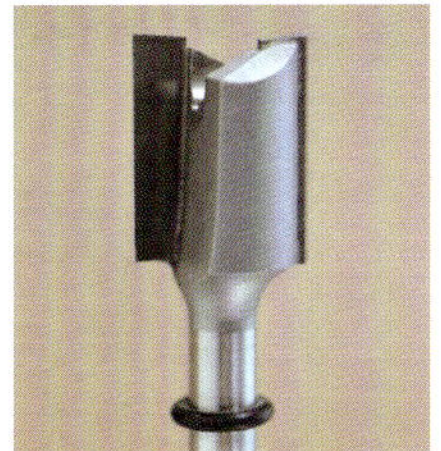

14

Processing the inset inro body. The bit height is set to the material thickness, from inset inro grooves, measured in step 6, to the outside edge. Once processing is complete, test fit the lid. If it's too tight, raise the bit slightly and continue to shave it down. The bit is a straight type with a slightly larger diameter.

15

Finally, adjust the height (length) of the inset inro using sandpaper on a flat surface, and it's complete.

Pegboard

In this section, we will learn about 'cope and stick joints' by making a pegboard. The stile and rail material has recesses and protrusions. This allows both the rail and stile tenon sections to be inserted into the recesses, creating a convenient joint. This technique can be applied to many projects, as seen in the cabinet doors below.

The photo below, on the right, we see an example of a cope and stick joint pegboard. Even though it was crafted with hand tools instead of a trimmer, the structure remains the same. The lid's tongue-and-groove joint is pretty much identical to the pegboard's cope and stick joint.

Here, we'll be using a trimmer table for processing. The key is to ensure that the groove is positioned at the centre of the material's thickness. No matter how carefully you set the fence to align the trimmer bit with the centre of the material, there will always be some discrepancies. Therefore, after the first pass, flip the material over and run the bit through the same groove again to remove these discrepancies. The groove width will be slightly wider, but it will ensure that the groove is accurately centred.

Once the groove is centred, subsequent tenon processing becomes much easier. If the tenon is also centred, the amount of material removed from both sides will be the same. Thus, after machining one side, you can simply flip the material over and machine the other side to achieve consistent thickness. If the tenon is not centred, you will have to adjust the height of the trimmer bit for each side separately.

In this section, we will use a ¼in (5.5mm) thick pegboard. Grooves must be cut to this thickness. The trimmer bit we will use is a ¼in (5mm) spiral bit. Thus, we will widen the groove using the previously mentioned method of processing from both sides twice to ensure the groove is centred in the material.

This cabinet is made using cope and stick joint techniques.

DIMENSIONAL DIAGRAM

476 (Sizing is Free)

Units: mm

35

Stile

Stile

FRONT SIDE

Rail

293 (Sizing is Free)

Groove Width: 5.5 and above

35

Rail

7

Rail

Scrap Material (For Test Cuts)

Stile

REAR SIDE

French Cleat on Frame Side

Wood Pegs

French Cleat on Wall Side

Wood Blocks for Adjusting Gap

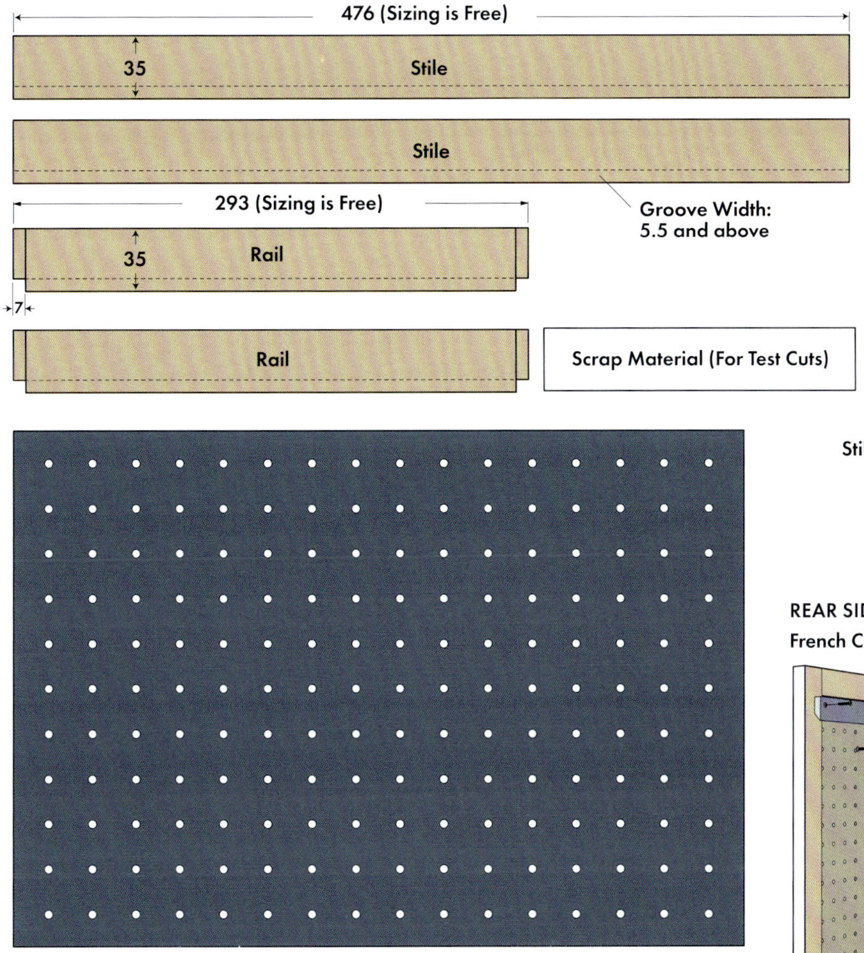

Pegboard (perforated board) ¼in (5.5mm) thick (size determined by dry-fitting and measuring internal dimensions).

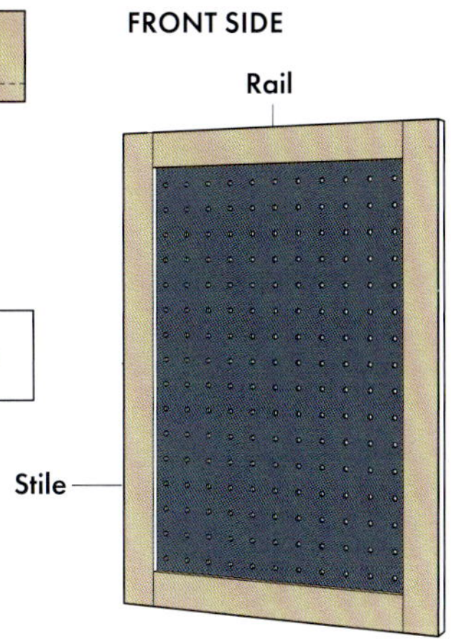

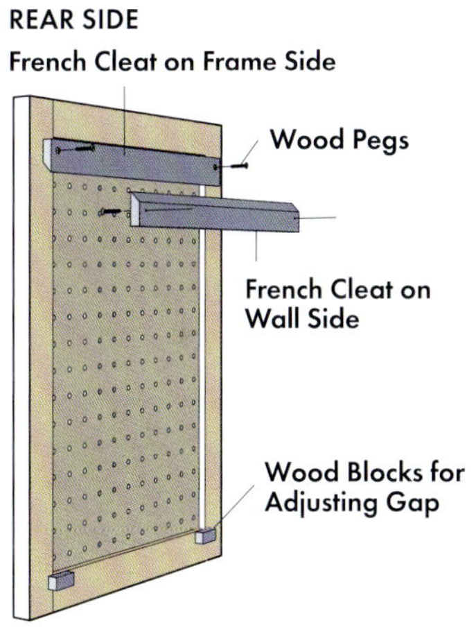

METHOD FOR PROCESSING CENTRELINE GROOVE

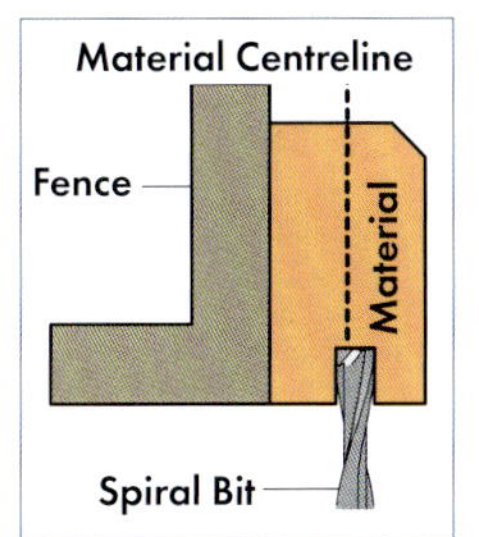

Centre bit on material and process groove.

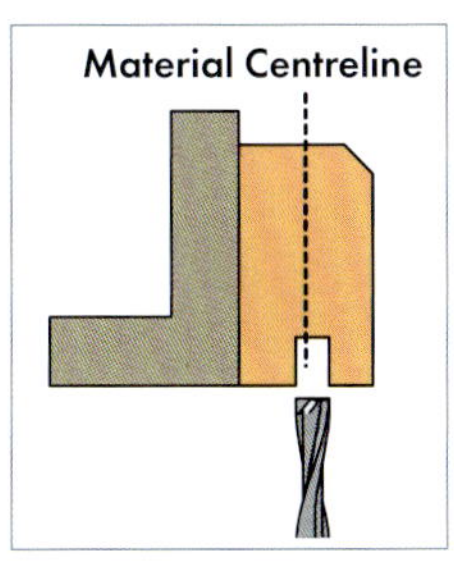

A perfectly centred cut isn't possible, it's off centre.

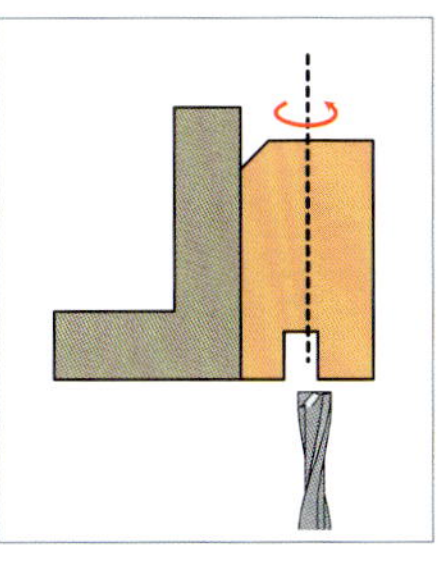

Flip material to make corrections (pay attention to angles).

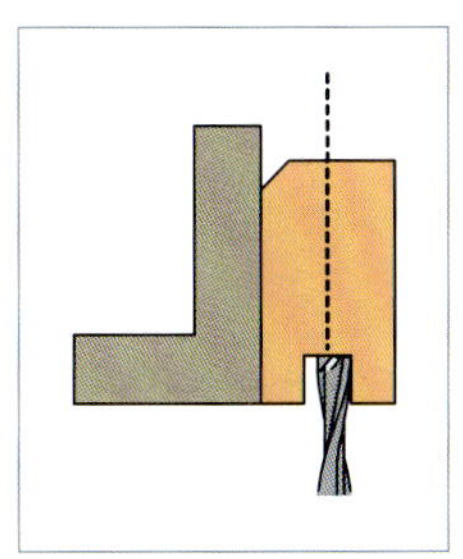

Pass bit through groove again to trim away any misalignment.

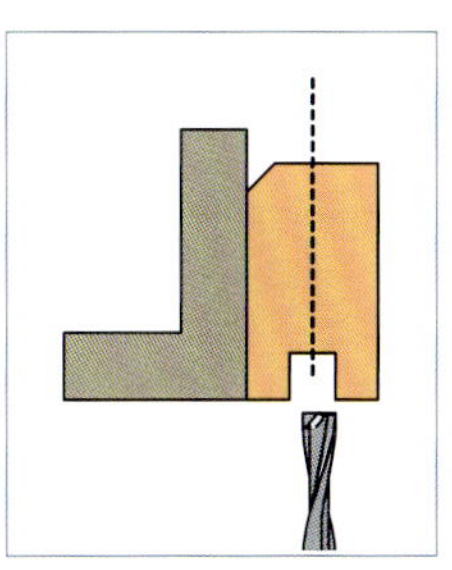

Even though the groove has widened slightly, we have still processed a centreline groove.

SET FENCE POSITION

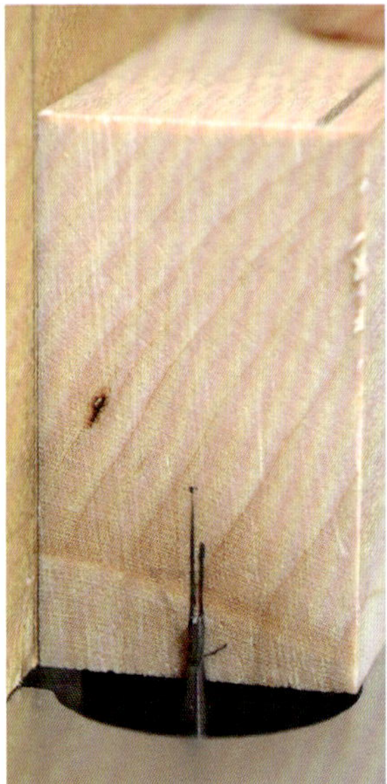

1

Measure the thickness of the material with callipers and set them to half that value. Then, using the callipers' depth gauge, draw a centre line on the material. Apply the callipers to the opposite side and draw another line.

2

Align the router bit visually with the mark made in step 1. It's easier to do if you use a thinner bit for alignment, as shown at left. Do a test cut. The groove width needs to be widened by 0.5mm, so intentionally offset the bit by 0.25mm when setting fence. This is a tiny adjustment, so rely on your senses. It's fine if the groove width ends up $\frac{3}{64}$in (1mm) wider than the panel thickness, as long as it's wider. When offsetting the bit to widen the groove, shift the fence away from the bit and secure. To avoid dangerous climb cutting, see 'Rules for Widening Groove Widths' on page 82.

GROOVE PROCESSING

3

Set the groove depth to $\frac{9}{32}$in (7mm) using an acrylic spacer. Set the trimmer hose clamp to this depth as well. This ensures that the maximum depth is set to $\frac{9}{32}$in (7mm).

4

Lower the bit. Make several passes to cut down to target depth. The initial depth should be around $\frac{5}{64}$–$\frac{1}{8}$in (2–3mm). For clarity in the upcoming steps, the letter 'A' is written on one end of the material and 'B' on the opposite end.

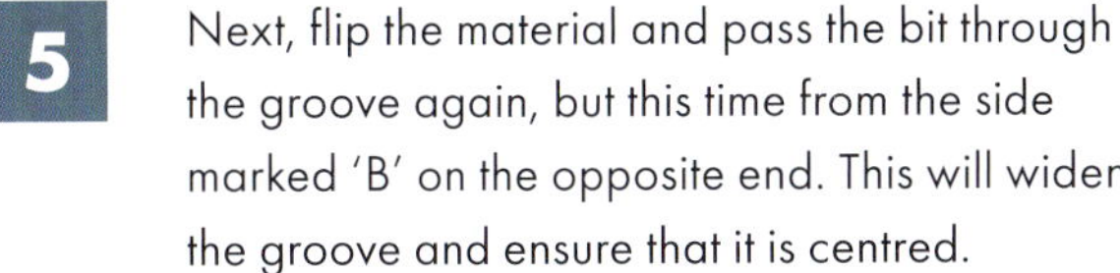

5 Next, flip the material and pass the bit through the groove again, but this time from the side marked 'B' on the opposite end. This will widen the groove and ensure that it is centred.

6 Repeat steps 4 and 5 while gradually raising the bit height. Process both rails and stiles. The short piece shown at the far right is excess material (used for test cuts).

7 Insert the panel into the groove. The groove width has been processed just right.

STILE TENON PROCESSING / TEMPORARY JIG FABRICATION

8 Create an H-shaped temporary jig that slides back and forth on the router table. Use scrap material for this.

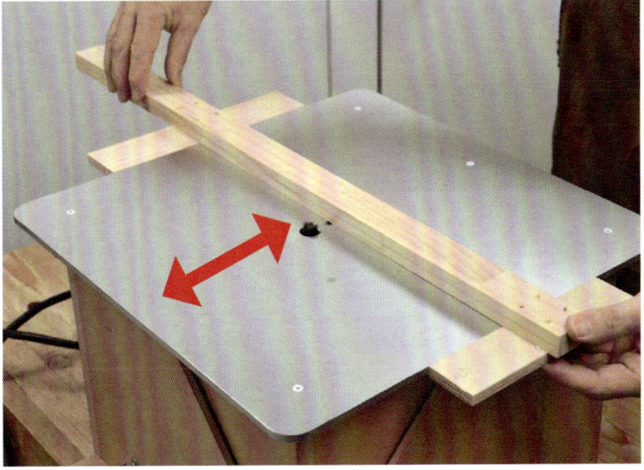

9 The jig is complete. It's used by sliding as indicated by the arrows.

SETTING TRIMMER BIT

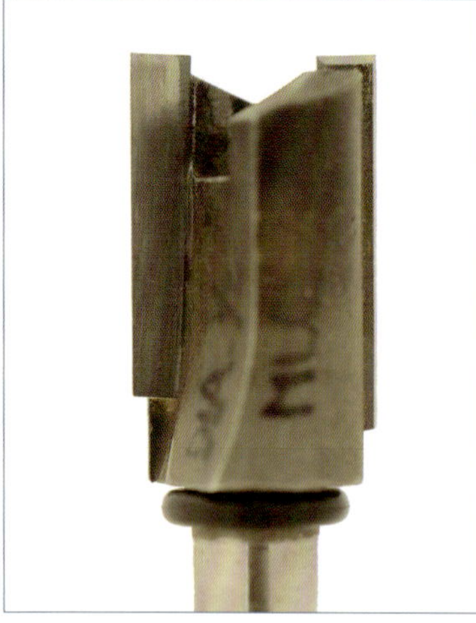

10

For tenon processing, you need a router bit with a diameter larger than the tenon length, as shown in the diagram. Here, a straight bit with a diameter of ½in (12.7mm) is used. The tenon length is 9⁄32in (7mm). Only the two horizontal rails will undergo tenon processing.

BIT FOR TENON PROCESSING

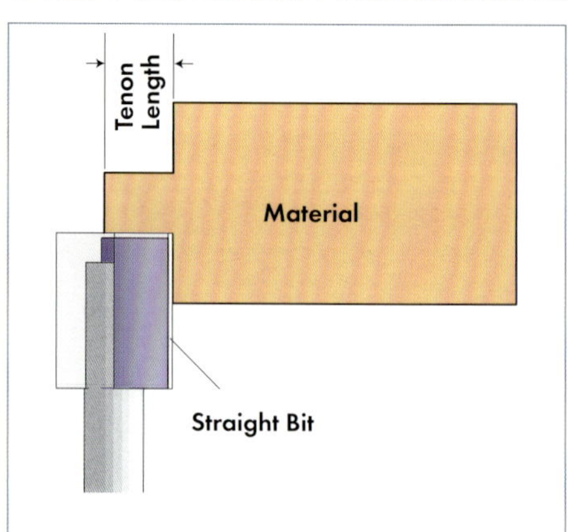

11 The groove width corresponds to the tenon thickness. Measure the non-grooved part thickness with callipers to determine how much to shave off both sides of the rail material. This value will be the bit height. Set the height using a spacer. Also, adjust the trimmer's hose clamp to this height.

12 The vertical pencil line marks 9⁄32in (7mm) tenon length. To cut up to this line with the trimmer bit, align the material accordingly. A stopper made from a small block is taped to the H-shaped jig with double-sided tape. This will ensure the material is positioned correctly for a 9⁄32in (7mm) tenon.

13 Gradually raise the bit height as you proceed. The masking tape seen at the lower right serves as a guide for the stopping jig.

14 After cutting one side, flip the material and process the other side.

15 Make the final cut at the set height. First, test with a scrap piece, check how well the tenon fits into groove, then cut the actual material.

16 This completes the tenon processing.

17 Here is the finished horizontal rail. The panel fits into the groove, and the tenon fits into the stile groove.

18 The fit is snug enough that it won't come apart when lifted.

19 If the fit is too tight, fine-tune it with a plane. In the photo, the tenon thickness is being slightly reduced using a shoulder plane. Sandpaper can also be used.

20

If the tenon is too long, adjust by slightly trimming the tip with a trim plane.

DRY FITTING

21

Dry-fit the assembly and measure the distance from groove to groove. Subtract $^{3}/_{64}$in (1mm) from that value and cut the pegboard accordingly. Apply water-based paint to one side of the board with a roller. Do not overload the roller with paint, as it can seep into the pegboard holes and affect its appearance.

22 After cutting the panel, dry-fit the assembly again and check for any joint gaps.

23 Since it will be difficult to finish the frame interior once assembled, finish only the inner side at this stage. In the photo, a Western-style hand plane is being used, but sandpaper can also be used.

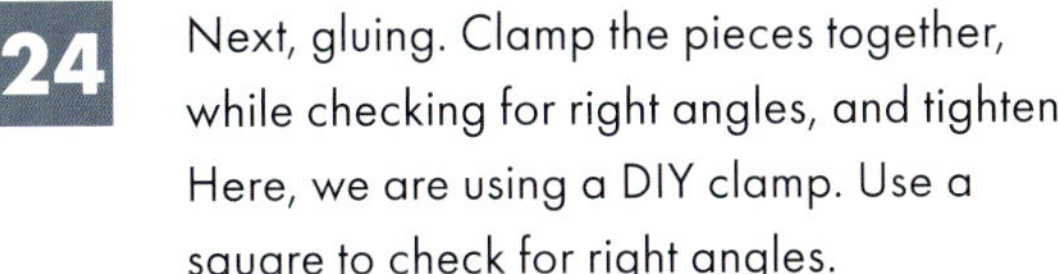

24 Next, gluing. Clamp the pieces together, while checking for right angles, and tighten. Here, we are using a DIY clamp. Use a square to check for right angles.

25 After the adhesive has dried, fix any slight joint irregularities with a plane or sandpaper.

26 Adjust any irregularities on the thick parts as well. After that, apply finish to the frame.

*** We used Asahipen Oil Based Gel Colour Varnish (White). The frame is maple. The panel finish is Nippe Home Products Water-based Wide V, Semi-gloss Feather Blue.**

◉ FRENCH CLEAT

To attach the pegboard to a wall, various hardware options are commercially available. However, in this case, we will make something called a 'French cleat'.

A French cleat, as shown in the diagram, involves cutting two pieces of wood at 45-degree angles and fitting their angled surfaces together to hang an item. This method can be used not only for pegboards, but also for hanging various objects. It provides a secure and firm attachment to the wall. Use this opportunity to experience what French cleats can do and apply that knowledge to your own projects.

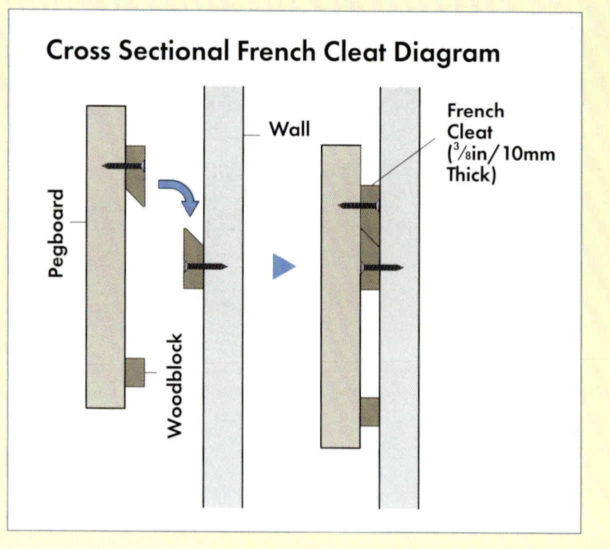

HOW TO MAKE FRENCH CLEATS

27

Use a 90-degree V-groove bit. First, extend the bit slightly beyond the material thickness. If the bit isn't large enough, you won't be able to process all of the material. Choose an appropriately sized bit; here, we are using a bit with a cutting diameter of approximately 1in (27mm). Process the material in several passes, gradually increasing the cut amount by moving the fence further away from the bit.

28

Process in several passes, leaving a flat area of about 0.5mm at the end. Angle both sides at 45-degrees, then cut in two pieces. This way, during routing, you can keep your fingers as far away from the bit as possible. For safety, process while the material is still wide.

29 Both pieces have been processed nicely.

30 Secure the frame to the back with wood screws. Take care to align the 45-degree edges.

31 The left cleat is the one that will be fixed to the wall.

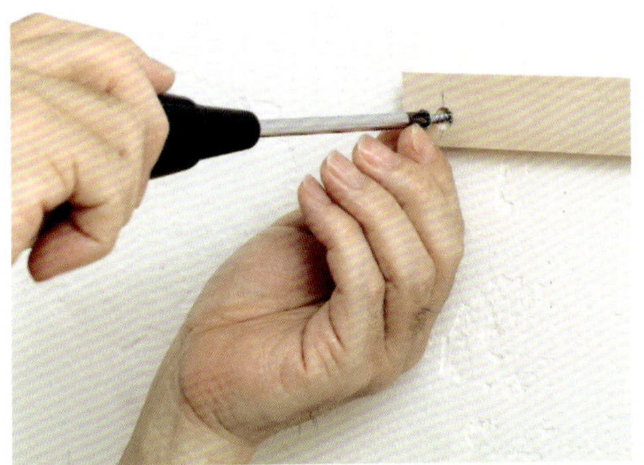

32 Once the wall-side cleat is installed horizontally, it will securely hold the frame just by hanging it from above.

Half-Blind Rabbet Joint Box

Let's create a box with a half-blind rabbet joint as a project to learn about rabbets and groove processing. We will process rabbets for this joint on the end panels (shorter boards) and create grooves for the base plate. The side panels (longer boards) will only have grooves for the bottom panel.

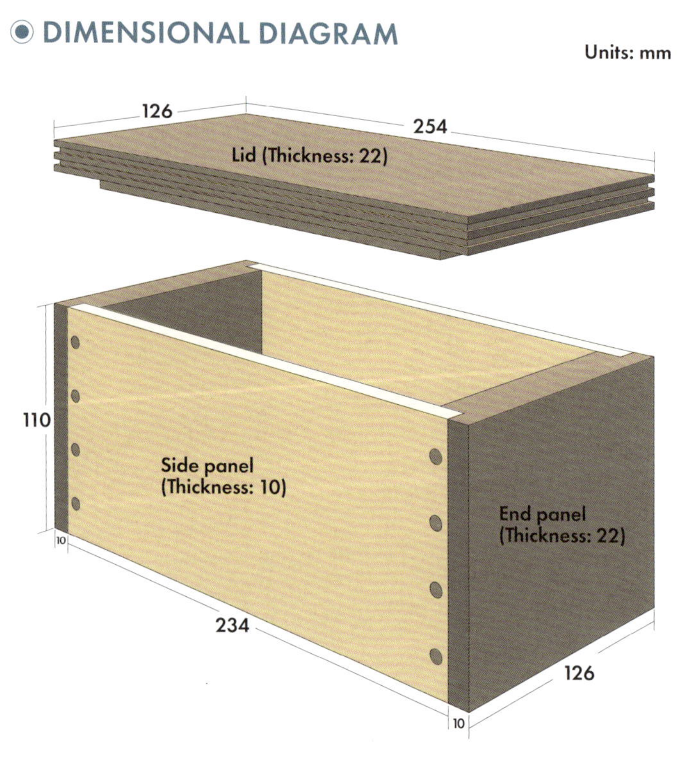

Cutting Plan

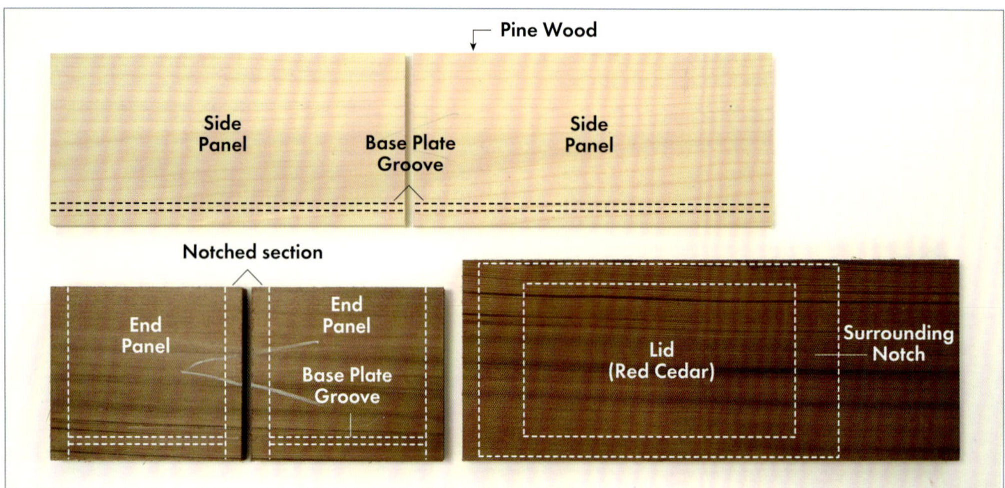

Side and end panels. Mark the boards with pencil or chalk in order to process the boards' bottom edges. The lid will be cut out after the box is assembled. In addition, you will need a base plate (5/32in/4mm thick) and a 1/4in (6mm) diameter round dowel.

Notch Depth

To cut the end panel out as shown in the diagram, set the distance from the fence to the bit. In this setup we will use scrap material from the side panel to account for its thickness. Place scrap material between the acrylic rod and the fence, as seen in step 2 on the opposite page, setting it so that the cutting edge touches the acrylic rod. While rotating the blade with your fingers, align it so that it contacts the acrylic rod at the correct radius. For your safety, always unplug the power cord from the outlet! Once the distance is set, secure the fence.

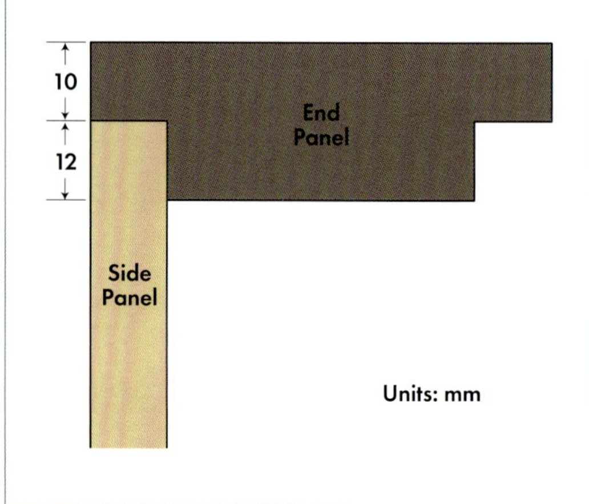

SAMPLE NOTCHES

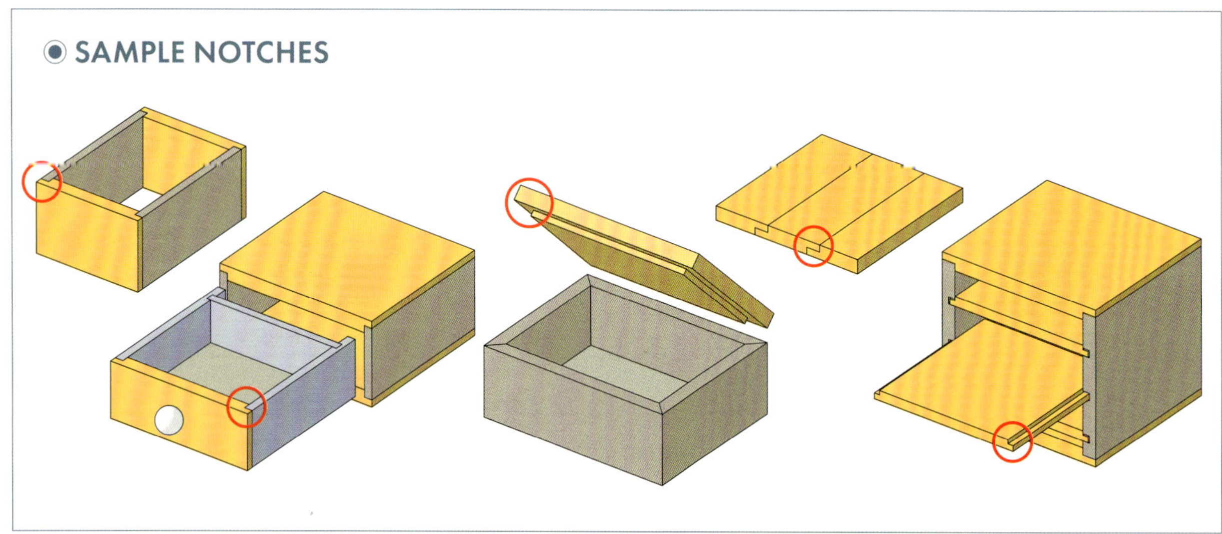

SETTING

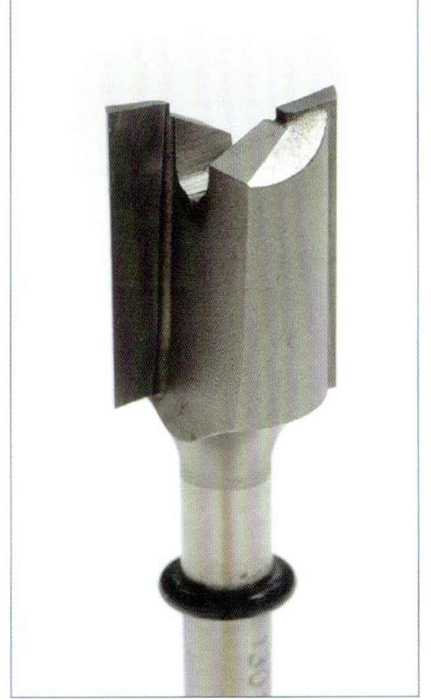

The straight bit shown in the photo is used. I chose a bit with a diameter of ½in (12.7mm).

2

Insert a piece of paper, about 0.5mm thick, between the scrap wood and acrylic rod. This will properly increase the cut. That way, when gluing, the end panel will slightly protrude from the side panel. Correcting this offset will result in a clean finish.

3

Next, determine the trimmer bit height. The maximum cutting depth is set to ½in (12mm). Place a depth gauge, set to ½in (12mm), on the table and touch the protruding trimmer bit from below. Then, secure the trimmer and also fix the hose clamp as a stopper. For details on the hose clamp, see page 37, 'Depth Setting With a Hose Clamp', and for the depth gauge, see page 57, 'Height Gauge/ Depth Gauge'.

SETTING

4 It's finally time to perform notch cutting on the end panels. We want to gradually cut to the desired depth, so we will lower the trimmer bit and start cutting when it is protruding just 5⁄64–1⁄8in (2–3mm) from the table (see page 83). Feed the material slowly. There's no need to press the material too hard against the fence. Be cautious not to let your fingers get too close to the fence! Keep them from sitting directly above the trimmer bit. If you have a dust collection system and a speed controller, use them. Cut notches on both end panels.

GROOVE PROCESSING

WHITESIDE Down-cut Spiral Bit, 1⁄8in (USA)

5

Now, we will process the base plate grooves. This is done on both the end and side panels. The bottom panel is 5⁄32in (4mm) thick, but the spiral bit is only 1⁄8in (3.2mm), so a single pass won't create a wide enough groove. There are two solutions: reduce the thickness around the bottom panel to match the groove width (notching) or widen the groove. This time, we will widen the groove. However, after the first cut, we'll shift the fence and perform a second cut. This can result in a climb cut if proper steps aren't taken, so check page 82, 'Rules for Widening Groove Widths'. Shift the fence away from the trimmer bit to widen the grooves. Process both the side and end panels. The groove depth is set to 1⁄4in (5mm). I will gradually remove material in several passes until the desired depth is achieved.

Groove processing a side panel. Fence is used on side with no bit opening.

Groove processing an end panel. Notching has already been performed.

BASE PLATE CUTTING/DRY FITTING/GLUING

6

To determine the base plate dimensions, we will dry fit the end panels and side panels. Rubber bands are useful for dry assembly. While in this state, we will measure the internal dimensions both vertically and horizontally, and add a total of ⅜in (10mm) for the groove depths on both sides to each measurement. If we use a measurement of 11⁄32in (9mm) when cutting the bottom panel, instead of ⅜in (10mm), there will be a tiny amount of 'play'. This ensures that the bottom panel fits securely. It also prevents gaps that may arise from the bottom panel being slightly larger than the joints of the rabbet joint. After checking the entire assembly for issues, glue it. Secure everything with two belt clamps and F-clamps. The bottom panel will not be glued but simply fitted in place.

Here we are checking the inner corner right angles. The four craft sticks at the joints are used as spacers. They will be pressed down later with four F-clamps.

BASE PLATE CUTTING/DRY FITTING/GLUING

7

Plane the entire assembly before moving on to the next step. Place the box on a flat workbench and lightly press down on all four corners; if it wobbles, there is a slight misalignment on the bottom between the end panels and the side panels, or excess glue. Correct this with a plane by gradually shaving the high parts off with diagonal cuts.

Aside from the method described in step 7, on page 129, there is also a correction method where you place sandpaper on a flat surface and sand the entire bottom of the box until it is level.

Applying water to the end grain makes planing a bit easier.

8

Additionally, the end panel's end grain is slightly higher than the side panel. This discrepancy should be corrected. The height difference is due to the 0.5mm thick piece of paper between the side panel and the acrylic square rod in step 2, page 127. Sanding end grain leads to a flat finish. If done in reverse, you would have to sand down the entire surface of the side panel.

REINFORCEMENT AND ACCENT DOWELS

The colour difference in the shavings indicates that the pilot holes have reached the end panel.

Drill pilot holes at the joints for ¼in (6mm) dowels. Dowels serve as reinforcement and also provide design accents. Check the lowest dowel position (leftmost dowel in the photo); it should be placed so that it doesn't collide with the base plate. The pilot hole size is 0.1mm larger than the dowel. Depth should allow dowel to enter the end panel by about 3/8–9/16in (10–15mm).

When dowels are pressed in, glue will be pushed out. If any comes out inside the box, wipe it away with a cotton swab or paper towel.

10

Apply glue to the pilot holes and push the dowels in using the handle of a mallet. As the dowels are pressed into the holes, glue will be pushed out. Additionally, glue will emerge from the end grain of dowels through wood fibres. Once the adhesive has set, trim the dowels using a flush cut saw. This helps ensure proper chip removal and makes cutting easier. If your saw has a set, it may damage the surrounding wood. (Note: 'Set' refers to slight outward splay of saw blade teeth.)

MAKING THE LID

Now, we can make the lid. The material is cut larger than needed and the edges are rabbeted with a trimmer. This rabbet will help stop the lid shifting. The rabbet is processed as shown in the diagram: first, rabbet both short edges and then rabbet both long edges.

The material is processed while still larger than the box. Begin by processing the short edges. As shown in the diagram, measure the long edge of the lid, then subtract internal dimension of the long board. Divide that result by 2 to get the rabbet depth on one side. This depth is the distance from the fence to the bit. However, setting this exact distance immediately can be dangerous. Even a small error could result in a gap, as the rectangular step could end up slightly too small. To avoid this, initially make the rectangular step slightly larger than needed, and gradually adjust it to the perfect size. To do this, slowly increase the distance between the fence and the bit. You can use spacers, just as when widening base plate groove (see page 80, 'Groove Widening').

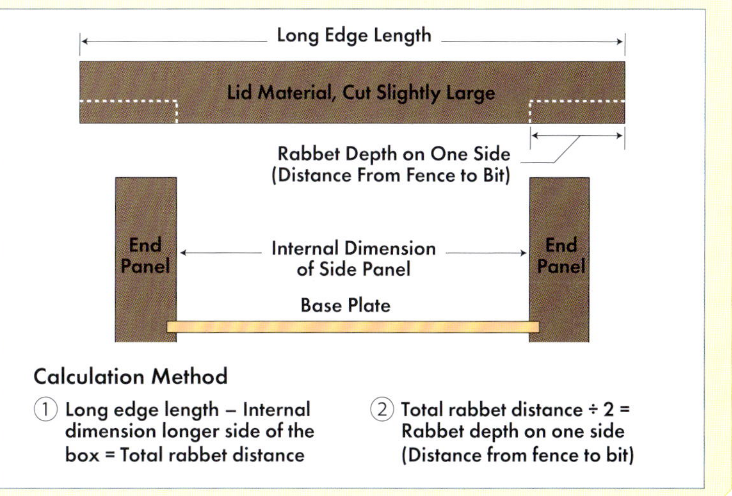

FENCE SETTING AND PROCESSING THE SHORT ENDS

12

Set the fence so that the rectangular step is approximately ⁵⁄₆₄in (2mm) larger on one side than the calculated size. The trimmer bit height is set to ¼in (5mm) using a depth gauge; the hose clamp serves as a stopper to maintain this setting for all lid notch cuts. The material will be cut in several passes, so adjust the bit height to ⁵⁄₆₄–⅛in (2–3mm) for the first pass. Process both short ends first. While rotating the bit by hand, fix the fence at a point where the bit diameter reaches its maximum. This operation is difficult, so make sure to set the distance within a safe margin and avoid making the rectangular notch too small.

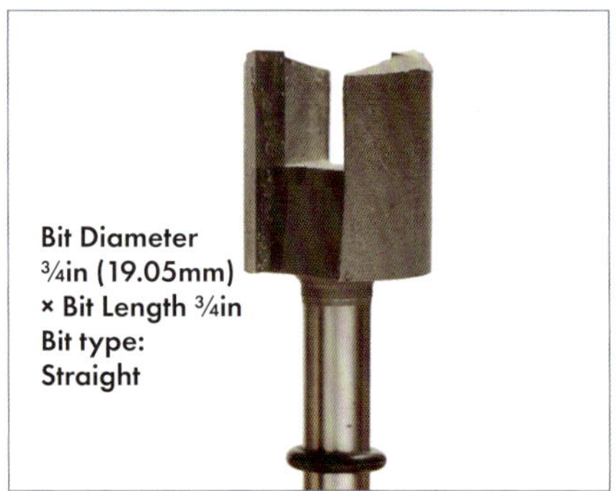

Bit Diameter ¾in (19.05mm) × Bit Length ¾in
Bit type: Straight

13

After repeating the process until the desired cutting depth is reached, dry fit. It's likely the lid won't fit the box yet. Use spacers to adjust the fence until the correct dimensions are achieved.

If splinters appear, carefully trim them with a cutter before proceeding with the second pass. This ensures a clean surface for the next stage and helps avoid further splintering or damage to the material.

PROCESSING THE LONG SIDES AND FINE ADJUSTMENTS

14

Process the long sides like the short sides. Here, final adjustments are made using a shoulder plane – a Western-style hand plane – but sandpaper can also be used.

The spacer for fine-tuning the fence adjustment is visible on the left side.

After completing the long side trimming, the fit may still be slightly tight.

The shoulder plane, which has a blade as wide as the body of the plane itself, allows trimming right up to the step edge. Use it to fine-tune the long sides to fit perfectly within the box.

Now the lid fits without any gaps and sits snugly in place.

TRIMMING THE LID PERIMETER

15

The lid is larger than the box, so the excess material around the edges needs to be trimmed off. If there is any play (any gap) between the lid and box, temporarily secure the lid to the box itself with double-sided tape. This ensures a tight fit before we carefully trim along the box edges with a saw. The saw used here is the Z-Saw 265 Asari Nashi (flush cut) blade. This is a flush cut saw so it won't damage the box surface. Its blade length of 10½in (265mm) is also very convenient. This specific blade is not available at typical hardware stores.

Carefully trim the lid edges by pressing the flush cut saw blade against the box as you cut.

For longer lid edges, where cutting isn't necessary, finish the lid using a hand plane.

LID GROOVE

16

Add sufficient spacers with the spiral bit protruding all the way.

With the spiral bit fully extended, insert a spacer for precise fence positioning. Adding two grooves to the lid's end grain and edge grain makes it easier to grip. A 1/8in (3.2mm) down-cut spiral bit is used. Cut depth is set to 1/8in (3mm), and the groove is made in two passes. Each pass cuts 1/16in (1.5mm) deep. When machining the perimeter, always begin with the end grain. This order ensures that any tear-out is cleaned up when cutting the next section, meaning the edge grain. Cutting sequence: end grain → edge grain → end grain → edge grain. (See 'The Right-Hand Rule' on page 22.) Lastly, round off the edges for each section and create a smooth finish.

Start on the end grain. Move the material slowly without applying too much pressure against the fence.

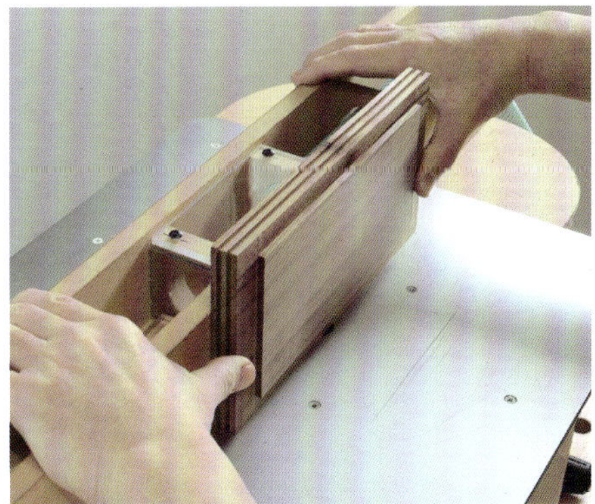

Finish with the edge grain. Grip both the material and the fence and slowly guide the workpiece to the left.

Wall-Mounted Shelf

Let's create a box with a half-blind rabbet joint as a project to learn about rabbets and groove processing. We will process rabbets for this joint on the end panels and create grooves for the base plate. The side panels will only have grooves for the bottom panel.

Process the dado joint grooves while the material is still in one long piece, meaning before it's cut into two side panels. This makes it easier to clamp the sliding fence to the material and secure the material to the workbench. Of course, you can also do this after cutting the material into two pieces. The bit used for this dado joint is a bearing-guided pattern bit, shown in the picture below. A shorter bit is ideal for this task.

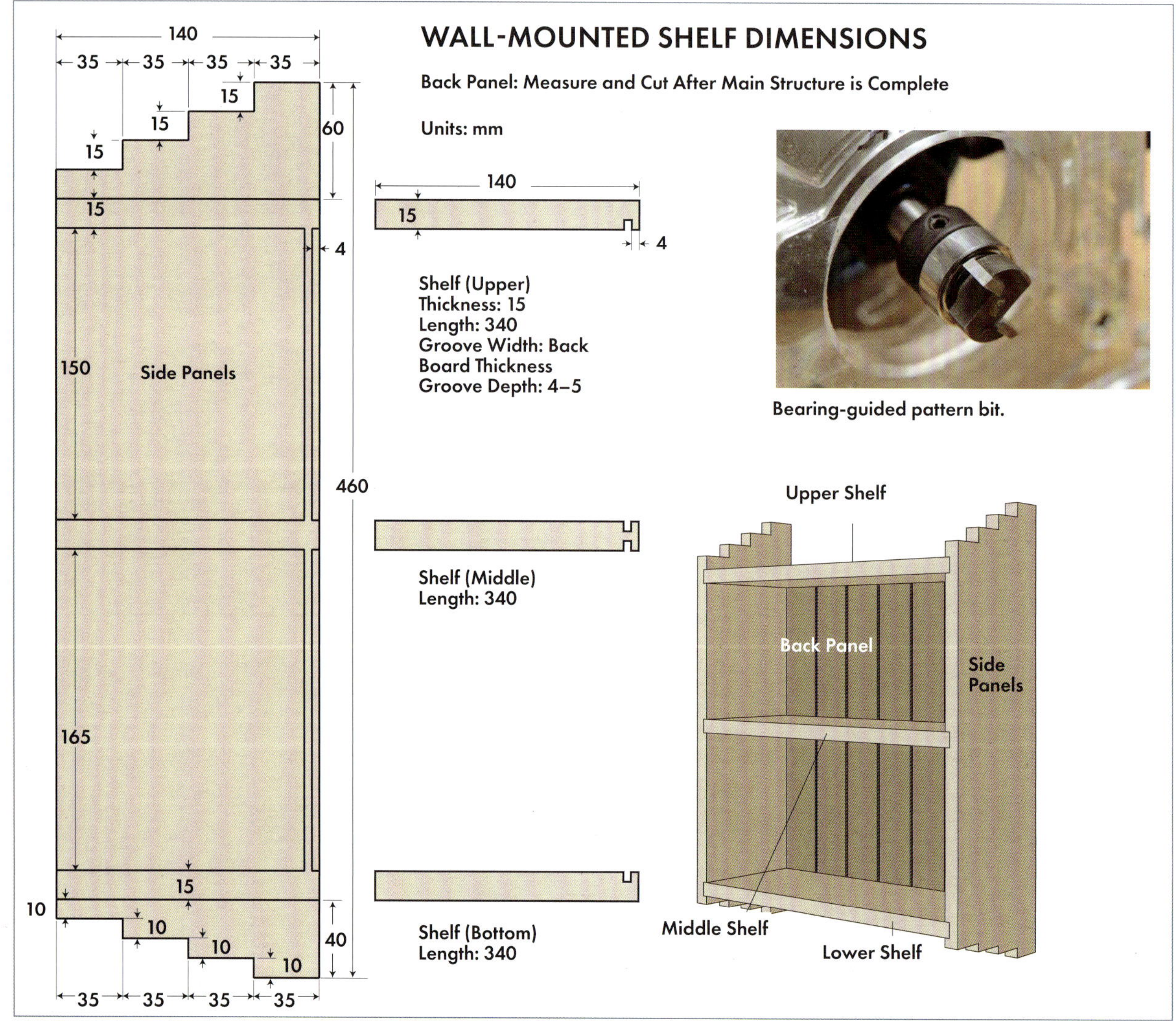

Bearing-guided pattern bit.

SETTING SLIDING FENCE TO SHELF BOARD THICKNESS

1 First, clamp the shelf board with the sliding fence and perform a test cut. If the fit is too tight and the shelf board cannot be inserted into the groove, adjust using the following method.

2 Insert a thin piece of paper, such as copy paper, along with the shelf board when setting up the sliding fence. This will slightly widen the groove.

PROCESSING DADO GROOVES / SETTING BIT DEPTH

3

Lower the trimmer bit to 5⁄32–¼in (4–5mm) below the sliding fence bottom to set the groove depth. Secure by lowering the hose clamp until it touches the transparent housing. Cut the groove in two passes.

4 Clamp the sliding fence and cut the groove into side panels. Be mindful of the cutting direction to avoid climb cutting. Remember the trimmer tends to 'curve left', so guide it along the left side of the fence (see page 20).

5 The dado groove is now complete. Use this method to cut all the grooves.

6 After finishing all the groove cuts, it's time to divide the side panels into two pieces.

CUTTING SIDE PANELS

7 Cut the side panels into two pieces.

DRY FIT

8 Perform a dry fit to check the alignment and fit of all parts.

◉ CORRECTING GROOVES

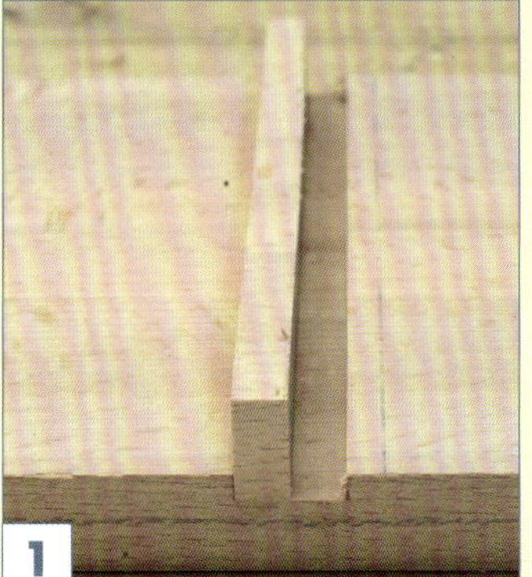

If a groove is cut wrong, it can be a hassle to buy new material and begin the process again. Instead, as shown in the picture, you can cut a short piece of scrap wood and glue it to one side of the already cut groove. Be sure to align the scrap grain direction with the main piece.

Trim any excess glued scrap with a saw. Then, use a plane or sandpaper to smooth it flat.

Continue smoothing until the surface is seamless to the touch. Edge grain (thickness part) will be corrected later.

Once the surface is smooth, re-cut the groove in the correct position. The repair is complete.

◉ ADJUSTING A SHELF IF IT DOESN'T FIT

If, during dry assembly, the shelf board doesn't fit into the groove, slightly shave the shelf's edge. Even if the fit seemed perfect during test cuts, it may be too tight during final assembly. This issue may seem like a problem with groove precision, but it's often caused by slight differences in board thickness before cutting. To correct this, attach a temporary fence to one side of the plane with double-sided tape and shave a taper of ¾–1⅛in (20–30mm). Adjust the height so that the temporary fence tilts the plane slightly to the left. However, the tilt is only about the thickness of a few sheets of paper, so the taper won't be noticeable to the naked eye.

PROCESSING THE BACK PANEL GROOVE

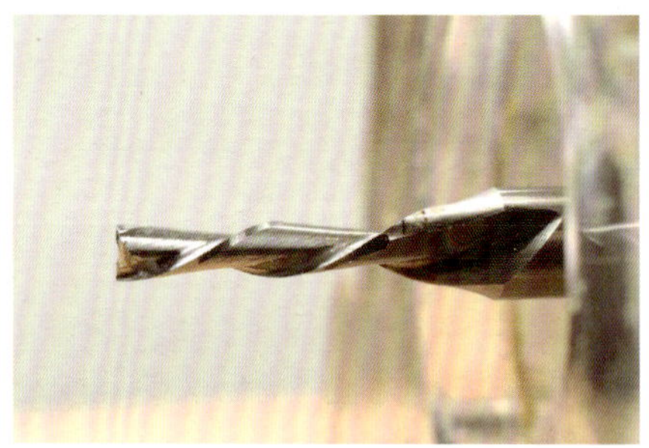

9 When dry fitting, mark the exact locations of the grooves with a pencil. The top and bottom shelves have grooves on one side only, while the middle shelf will have grooves on both sides (see the dimensional drawing on page 136).

10 The back panel is made of 5/32in (4mm) Japanese linden plywood. The bit diameter is 1/8in (3.2mm). One pass won't make the groove wide enough, you'll need to cut twice. A down-cut spiral bit is used. This helps minimize groove burrs.

SETTING THE TRIMMER BIT DEPTH

11 Place a transparent spacer on top of the base material and set the trimmer on top. This spacer determines the bit cutting depth. Once the depth is set using the black lever, lower the hose clamp to lock it in place and use it as a stopper.

12 Attach a straight guide to the trimmer and, once the position is set, cut the groove. Also, cut a groove in a piece of scrap wood now, to use later for test cuts when widening the groove.

13 Use scrap wood to determine how much to adjust the groove width. Then, proceed with the final cuts. You'll notice a slight widening of the groove.

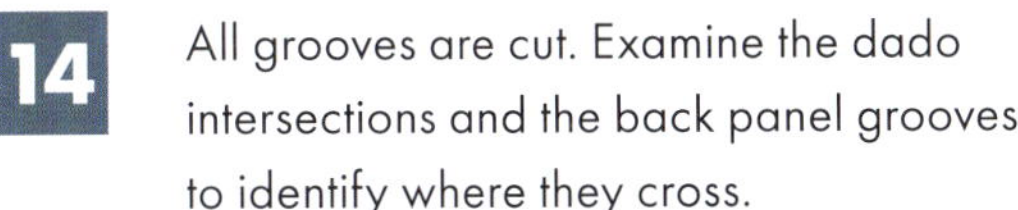

14 All grooves are cut. Examine the dado intersections and the back panel grooves to identify where they cross.

15 Dry assemble to ensure the groove positions align correctly.

BACK PANEL

16

The back panel is cut from 5⁄32in (4mm) plywood that has thin vertical grooves to give the appearance of planks. A straight guide is attached to the trimmer for this process. As shown in the picture, it is efficient to cut grooves on both sides with the same setup.

17 As you move towards the back panel's centre, you'll need to extend the straight guide arm. However, there's a limit to how far it can reach. To address this, remove the straight guide, leaving only the arm. Then, clamp a piece of scrap wood to it as a temporary fence. This will help gain the necessary distance to cutting the second groove.

18 Apply double-sided tape for additional security while the temporary fence is still attached.

19

These are the last groove cuts. Even with the temporary fence, this is the maximum distance it can reach.

COLOURING THE BACK PANEL

20

The main body is finished with clear lacquer. The colour is expected to deepen, so we'll apply a dye to the light-coloured birch plywood to match it. It's a powder type that dissolves in hot water, purchased from an international online store. Once dry, the back panel is cut into two pieces, one for the top and one for the bottom.

DECORATIVE SIDE TREATMENT

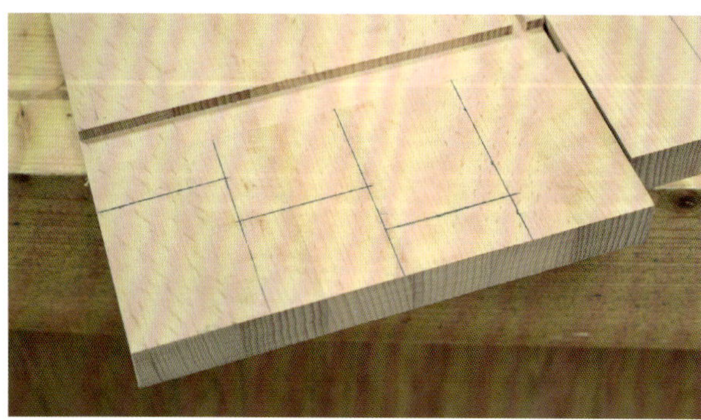

21

Draw pencil lines for step cutting the top and bottom panels.

22

Make vertical cuts with a jigsaw.

23 Next, make horizontal cuts to create the stepped shape.

24 Sandpaper off any burrs. I'm using a sanding block of aluminium square pipe with sandpaper wrapped around it and double-sided taped. It's handy because all four sides can be used. This idea came from a member of a woodworking club.

25 Insert two back panels and dry fit.

GLUING

26 Apply glue to the dado grooves and assemble the entire piece.

27 Tighten the clamps on each shelf, one at each end. Wipe off any excess glue. Use a square to check the interior. If it's not square, adjust by applying diagonal pressure with a clamp (see page 61).

FINISHING

28 Clean up any unevenness with sandpaper or a plane.

29 Chamfer the edges and clean any shavings or dust.

PAINTING

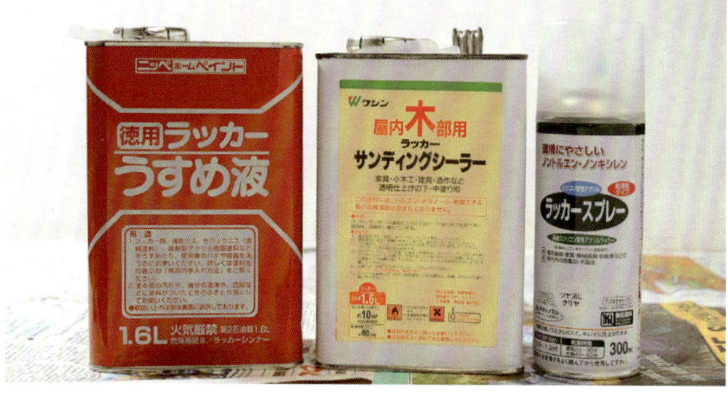

30 I often apply sanding sealer, sand that down, and then finish with a matte clear lacquer. The matte finish produces a calm, subdued look.

31 Here, we are soaking the brush we previously used to apply sanding sealer in lacquer thinner because it had become stiff. Plastic food bags are very useful for this. Even if you're painting over several days, you can seal the brush in the bag with a zipper and it will remain usable.

32 Apply sanding sealer as a base coat.

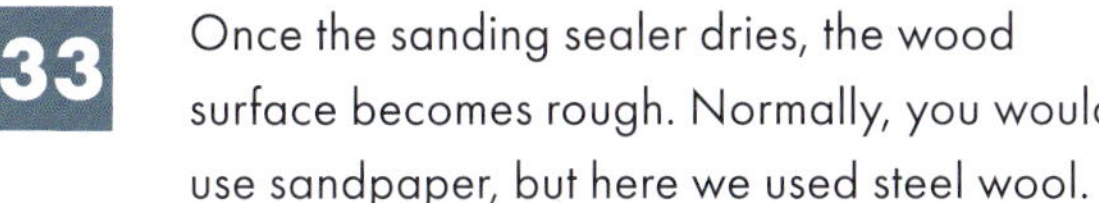

33 Once the sanding sealer dries, the wood surface becomes rough. Normally, you would use sandpaper, but here we used steel wool.

34 I created a simple spray booth using cardboard as a screen and sprayed on the matte clear lacquer.

35 To make it easier to spray along the wood grain, we positioned the piece horizontally. It's on a store-bought turntable so that it can be rotated when sprayed, thus ensuring no missed spots.

36 The piece has been nicely coated, giving it a calm, refined look. Once the wall-mounting hardware is attached, it's complete.

Constructing a Box Joint Jig

The common box joint is frequently used in Japanese masu (box-like measuring cups that store rice and sake). It's a very sturdy joint that is also used in high-end cabinets.

Let's make a jig for cutting these box joints, with a special feature that allows us to adjust the fit. The trick to cutting proper notches involves using a straight or spiral bit to cut recesses, and then repeatedly inserting the jig's key into consecutive sockets and cutting again.

The box jig's dimensions are designed for use with the MIRAI trimmer table set. Additionally, the trimmer uses a spiral bit of ¼in (6.35mm), for both cutting diameter and shank diameter.

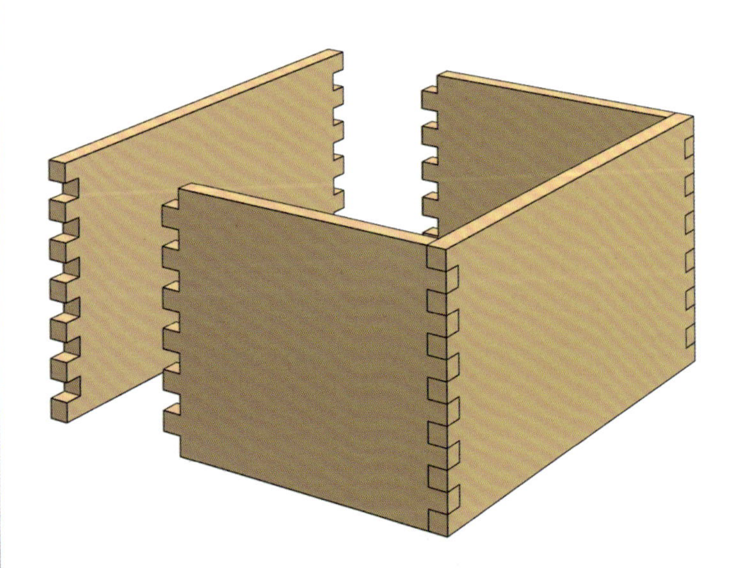

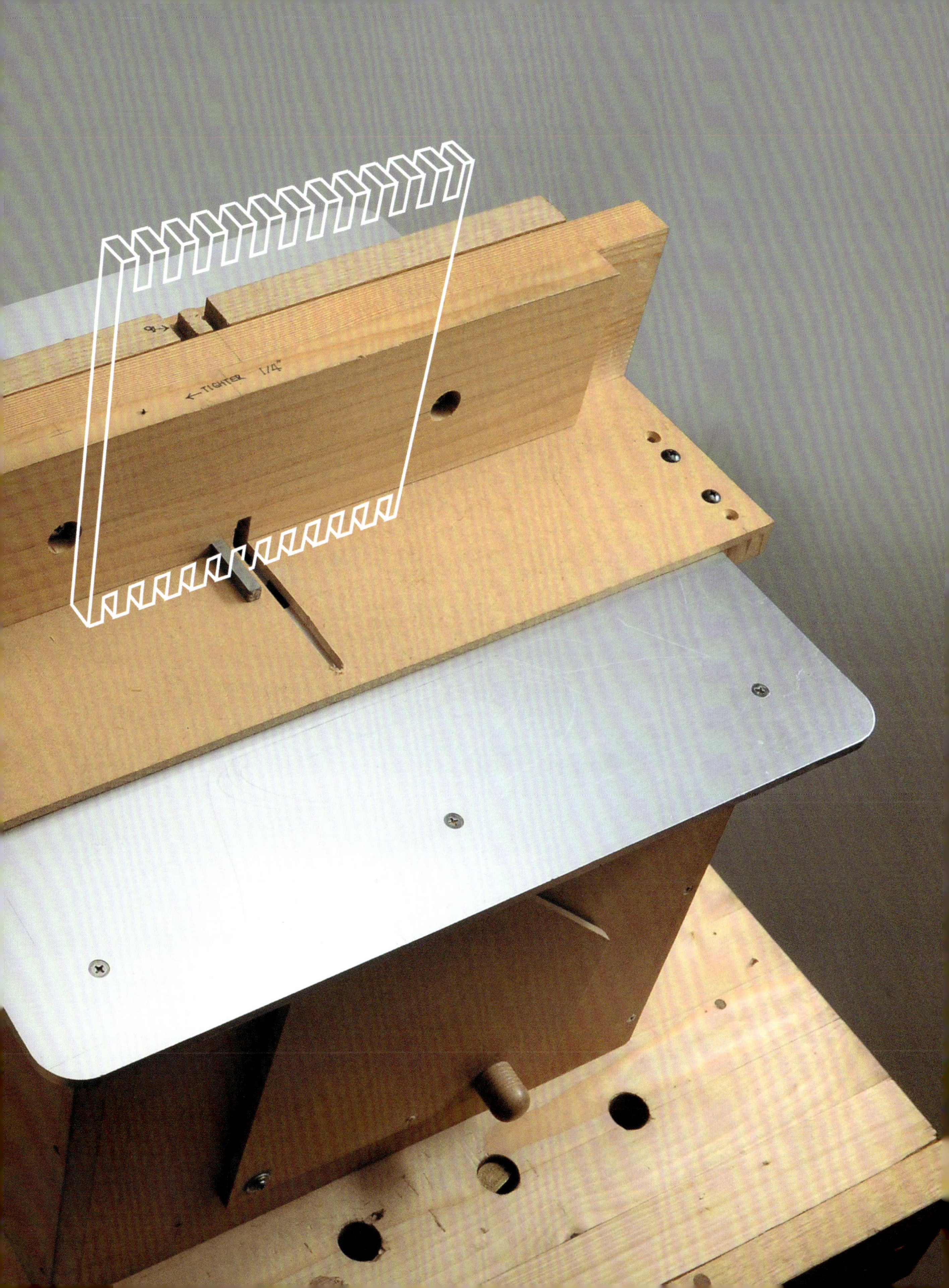
←TIGHTER 1/4"

BOX JIG CUTTING DIAGRAM AND SCREW POSITIONS

Units: mm/t: thickness

450
Centreline
Fixed Fence
t: 18–20 Laminated Board
80
400
12
Centreline
Movable Fence
t: 18–20 Laminated Board
70
12mm Line
450~460
Centreline
75
110
70
Centreline
75
240
Base Plate (t: 5 MDF) and Screw Positions
25
260
Key Material (Made From Hardwood)
Width: Bit Cutting Diameter
Length: Approx. 100mm
(cut down to 45mm for key material, scrap will also be used)
Thickness: 5
Stopper (2 Pieces)
25×260×15 Square

BOX JOINT JIG STRUCTURAL DIAGRAM

Units: mm/t: thickness

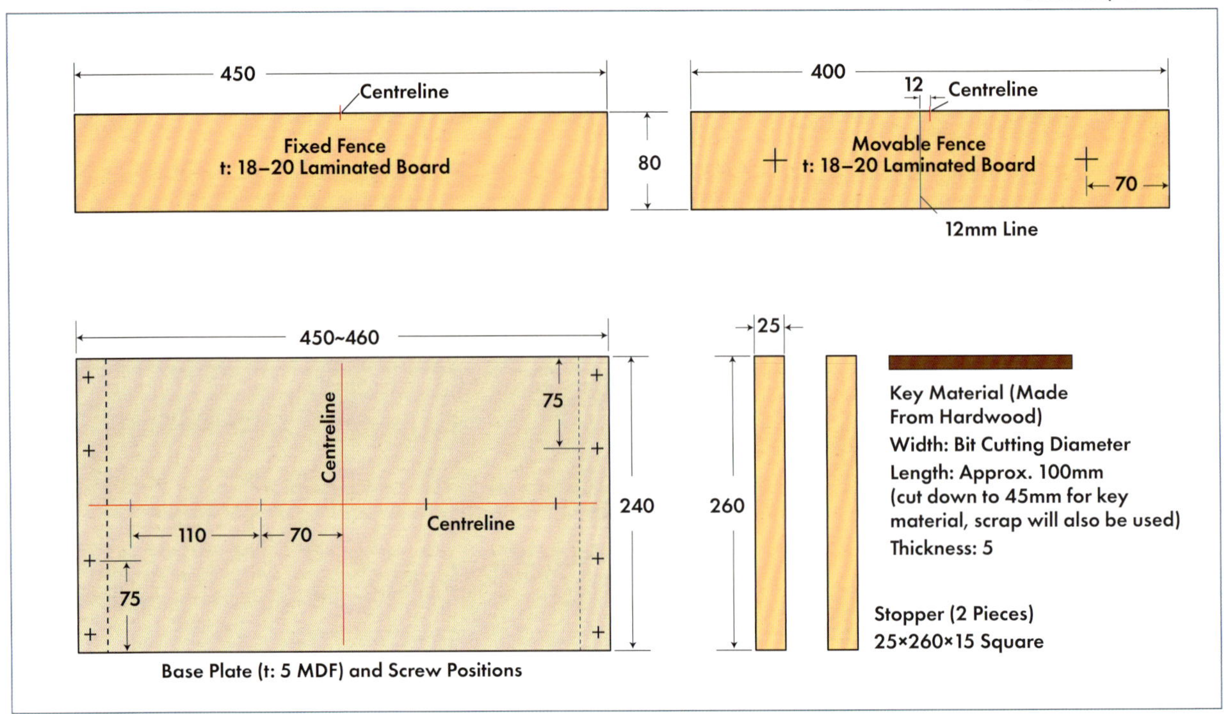

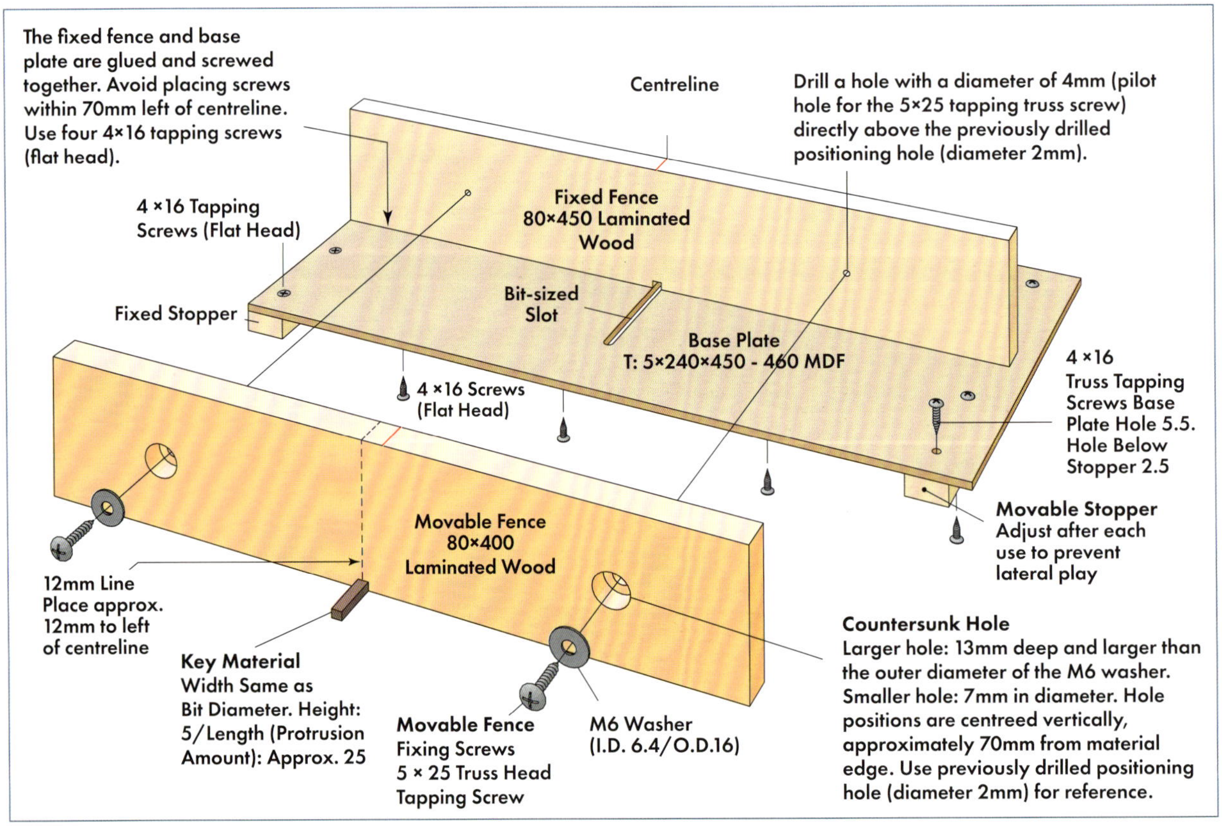

The fixed fence and base plate are glued and screwed together. Avoid placing screws within 70mm left of centreline. Use four 4×16 tapping screws (flat head).
Centreline
Drill a hole with a diameter of 4mm (pilot hole for the 5×25 tapping truss screw) directly above the previously drilled positioning hole (diameter 2mm).
4 ×16 Tapping Screws (Flat Head)
Fixed Fence
80×450 Laminated Wood
Fixed Stopper
Bit-sized Slot
Base Plate
T: 5×240×450 - 460 MDF
4 ×16 Screws (Flat Head)
4 ×16 Truss Tapping Screws Base Plate Hole 5.5. Hole Below Stopper 2.5
Movable Stopper
Adjust after each use to prevent lateral play
Movable Fence
80×400
Laminated Wood
12mm Line
Place approx. 12mm to left of centreline
Key Material
Width Same as Bit Diameter. Height: 5/Length (Protrusion Amount): Approx. 25
Movable Fence Fixing Screws
5 × 25 Truss Head Tapping Screw
M6 Washer
(I.D. 6.4/O.D.16)
Countersunk Hole
Larger hole: 13mm deep and larger than the outer diameter of the M6 washer. Smaller hole: 7mm in diameter. Hole positions are centred vertically, approximately 70mm from material edge. Use previously drilled positioning hole (diameter 2mm) for reference.

◉ BOX JOINT JIG STEPS

Units: mm/t: thickness

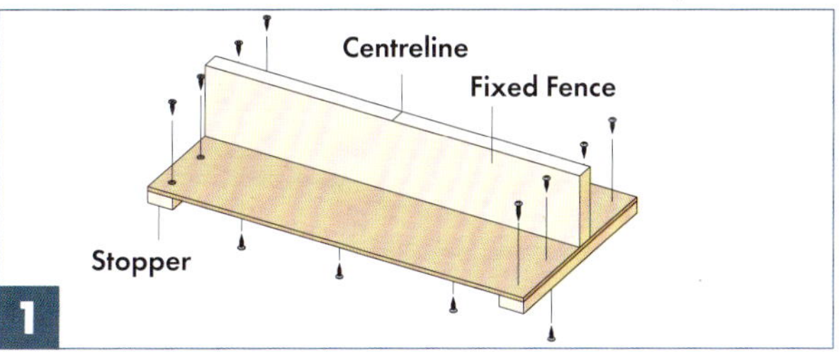

Constructing Basic Jig Form

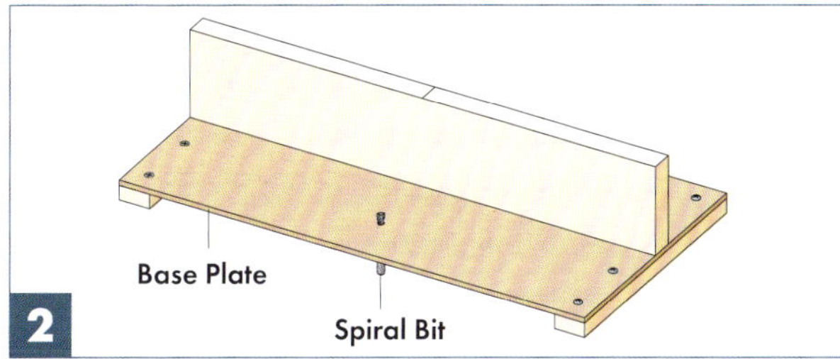

Extend the spiral bit slightly above the trimmer table and turn on the trimmer. Lower the jig over the bit to drill a hole into the base plate. Extend the spiral bit upward if needed.

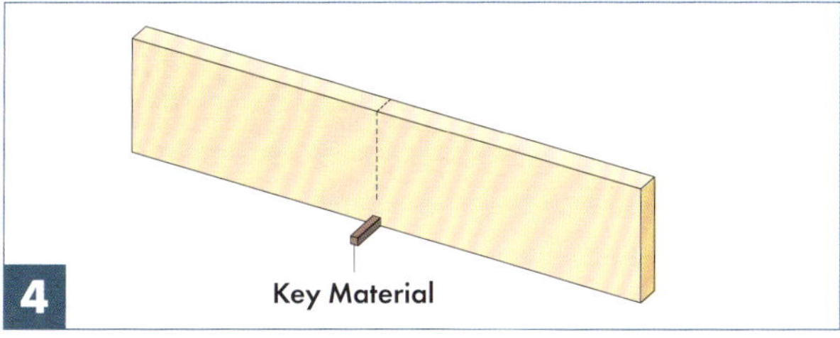

Fit the key material into the movable fence cut-out and glue in place. The key material width should match the spiral bit diameter. Use hardwood for the key. The protrusion from the movable fence should be about 1in (25mm).

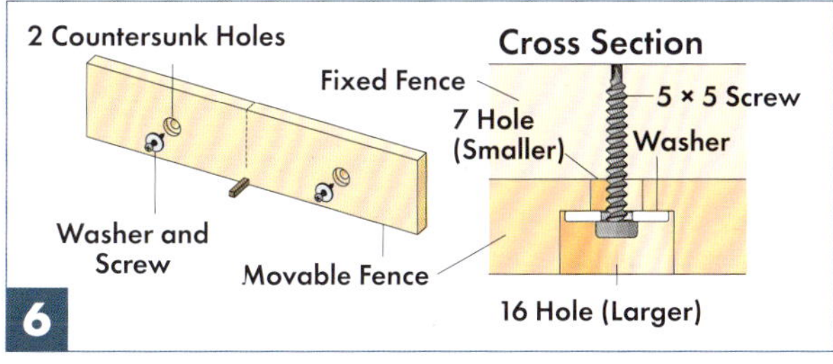

Loosening screws that hold the movable fence will intentionally create play to allow for adjustments. Drill countersunk holes in the movable fence. The larger hole is for the washer. The smaller hole should have a diameter of 9⁄32in (7mm). The 5⁄64in (2mm) hole in the fixed fence should be enlarged to 5⁄32in (4mm).

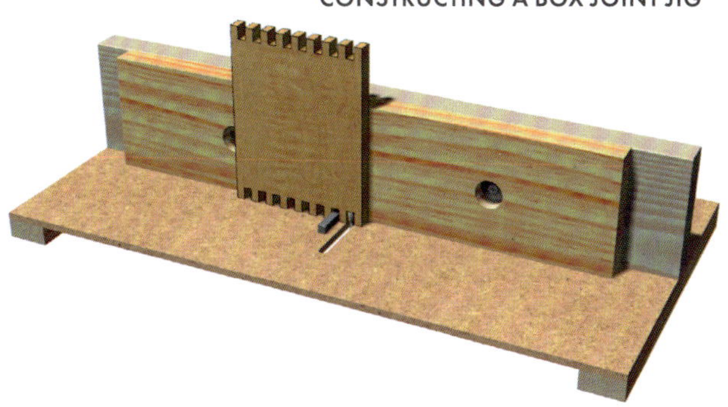

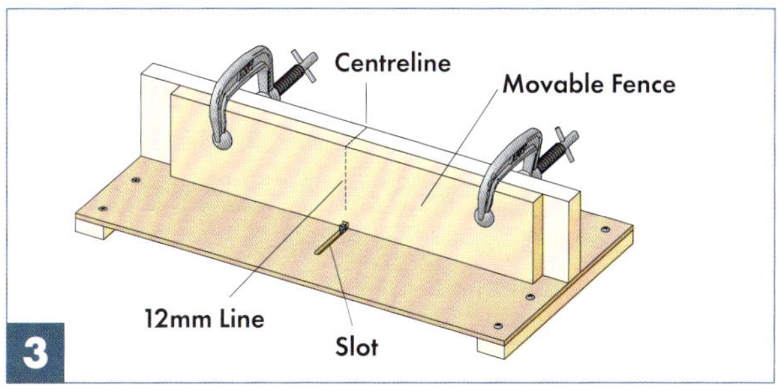

Clamp the movable fence to fixed fence while aligning the centreline and ½in (12mm) line. Move the jig to create a slot in the base plate. Cut into both fences. Set the spiral bit height to be exactly the same as the key material.

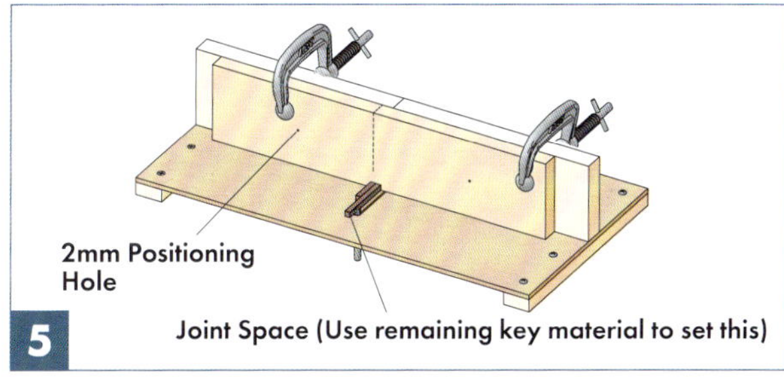

Set and clamp the movable fence so that the space between the key material and spiral bit matches the bit diameter. Use the remaining key material for this. We will refer to this as the 'joint space'. Drill 5⁄64in (2mm) positioning holes in both fences. Next, disassemble the two fences and drill different sized holes in each.

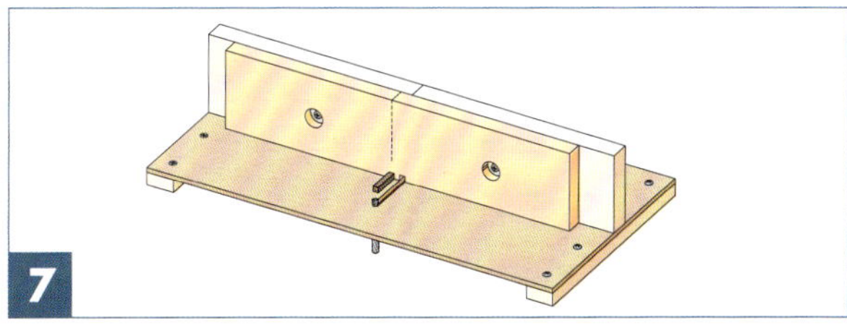

The box joint jig is now complete. Narrowing the joint space loosens the fit, while widening it tightens the fit. It's advisable to perform tests with scrap material before beginning serious operations.

CONSTRUCTING THE BASE

1 Attach a fixed stopper to the left end of the base plate. Apply instant glue to the stopper and use a square to bond it at a right angle. After the adhesive sets, secure with screws.

2 Secure the base plate and fixed stopper with four 4×16 tapping screws (flat head). The material should not shift as instant glue has already been applied. Countersink pilot holes, and ensure the screw heads are slightly recessed below the base plate surface.

INSTALLING THE FIXED FENCE

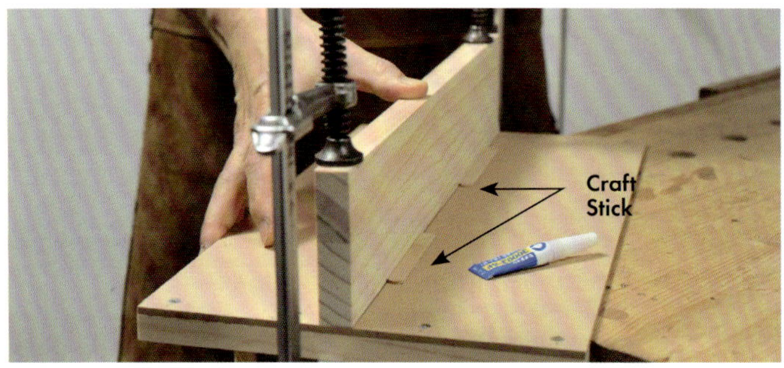

3 Draw a centreline for positioning. Attach the fixed fence to the centre of the base plate with instant glue, and clamp both ends. If possible, attach craft sticks, with double-sided tape along the centreline. This will make it easier to position the fixed fence. Ensure the fence is perpendicular to the base plate by checking before screwing.

4 Secure the fixed fence with four 4×16 tapping screws from the underside of the base plate. Be careful not to place any screws where the trimmer bit will pass.

◉ REQUIRED SCREWS

* 4×16 truss head tapping screws: 4 pieces (for movable stopper)
* 4×16 tapping screws (flat head): 8 pieces (4 for tight stopper and 4 for fixed fence)
* 5×25 truss head tapping screws: 2 pieces (for movable fence)
* M6 washers: 2 pieces (outer diameter: ⅝in/16mm for movable fence)

MOVABLE STOPPER

5 The movable stopper is used to eliminate lateral play (looseness) between the trimmer table and jig during each use (temperature and humidity can cause lateral play). Temporarily secure the movable stopper with double-sided tape. Ensure to sandwich the trimmer table edges between the already installed tight stopper and the movable stopper.

6 Drill positioning holes for screws with a 5⁄64in (2mm) bit in the stopper. To ensure that the holes are drilled vertically, use an acrylic mirror with holes drilled in it to align the reflected drill bit and the actual bit.

7 Widen the base plate hole diameter to ¼ (5.5mm). The stopper hole diameter will be 3⁄32in (2.5mm).

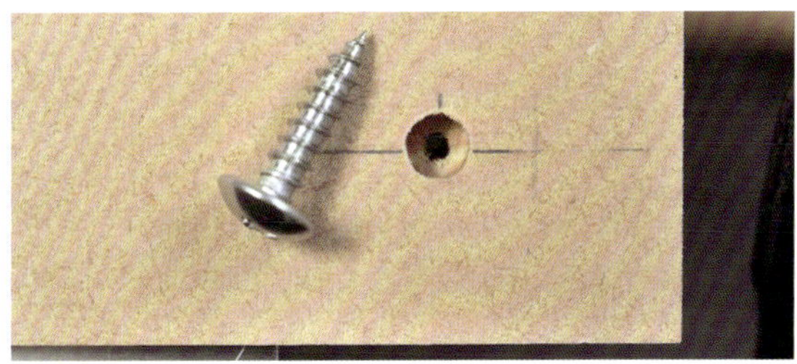

8 When the two holes overlap, as shown in the photo, the movable stopper will have a little play and movement. This helps eliminate slap (large movement) that could occur between the trimmer table and the jig.

OPENING SLOT

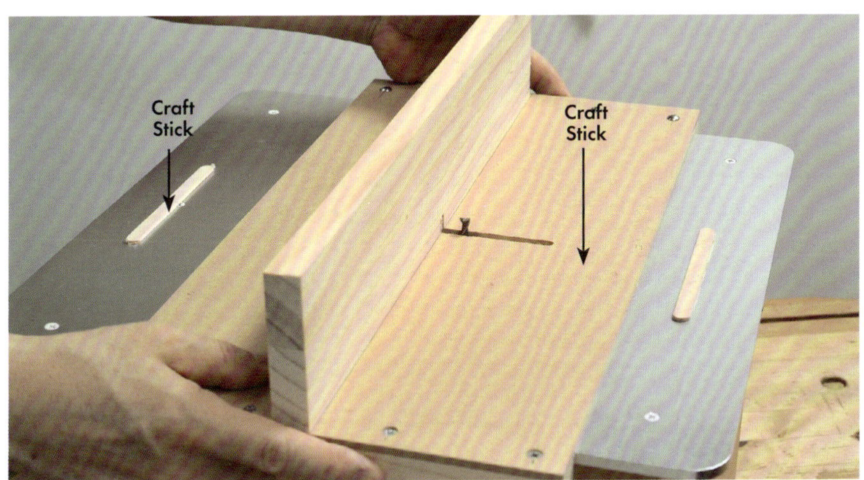

9 Place the box joint jig on the trimmer table and position it over the rotating bit to drill a slot along the centreline. The bit height should be approximately 3⁄8in (10mm). Attach craft sticks with double-sided tape on each side of the table to limit travel distance. The slot length should be 15⁄8–2in (40–50mm). Drill until the bit is concealed within the fixed fence.

KEY MATERIAL INSTALLATION

10

Prepare the key material to the same width as the bit diameter. Create a groove in the movable fence and attach the key material there. Adjust the bit to match the key material thickness. The trick is to raise the bit by a single piece of paper. Next, place the movable fence in front of the fixed fence. Clamp it so that the fixed fence centreline aligns with the ½in (12mm) line on the movable fence. Create a groove, and insert the key material.

11

Next, the key width material will be reduced to ¼in (6.35mm), the same as the spiral bit. Use a plane (I used a block plane) or sand the key down. Attach key material to a scrap base plate with double-sided tape. Finish while checking with callipers.

12

Insert 1¾in (45mm)-long key material into the groove, apply glue, and clamp in place. Ensure the key material and fence are flush, without any offset. Protrusion length from the movable fence should be about 1in (25mm).

MOVABLE FENCE INSTALLATION AND HOLE DRILLING

13

This task involves drilling holes in the movable fence so that it can be attached to the fixed fence. Use a scrap piece of key material as a gauge to ensure that the space between the key material and bit matches the bit diameter (should match key width). Rotate the spiral bit, positioning it so that it contacts the end of the key material.

14

Once the position is determined, secure with clamps. Drill the positioning holes as described in step 13.

15

Use a 5/64in (2mm) drill bit to create through holes in both fences. Drill a positioning hole 2¾in (70mm) from the end of the movable fence (see page 148). After that, disassemble the two pieces and drill different sized holes in each one.

COUNTERSUNK HOLES IN MOVABLE FENCE

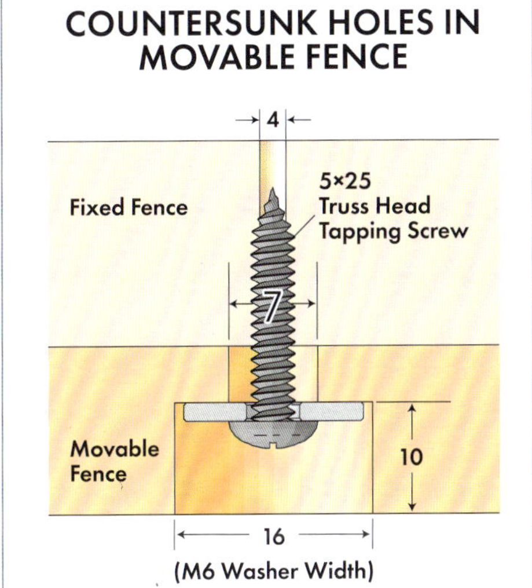

16 Countersink holes in movable fence. First, drill a larger hole 5/8in (16mm) to accommodate an M6 washer. Then, drill the smaller hole 9/32in (7mm). Widen the holes in fixed fence to 5/32in (4mm).

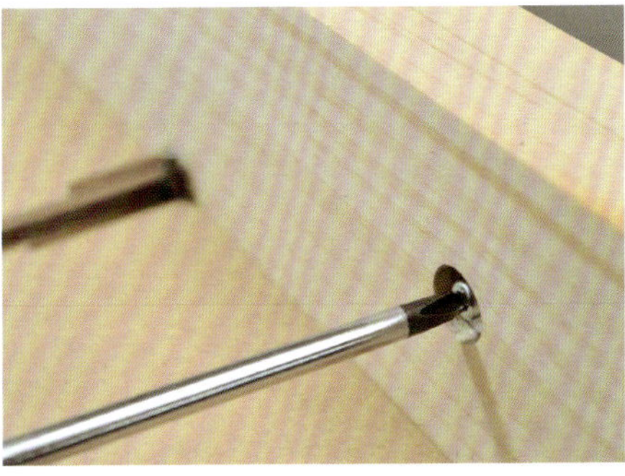

17 Once the holes are drilled, assemble the unit. Secure the two fences with washers and screws while placing a scrap piece of key material between the fixed key material and bit. This completes the assembly.

Box Joint Box

Let's try to make a basic box joint model. Creating box joints is incredibly rewarding. There is so much joy in mastering a seemingly difficult joint technique on your own.

The basic principle here is to use a 'progressive feed', cutting and then moving the piece for the next cut. A spiral bit (or straight bit) can be used to produce both the box joint pins and sockets. Inserting the 'key' into the socket portion – created by the initial spiral bit cut – allows us to accurately and consistently cut each successive socket.

Perform a test cut on a piece of scrap wood and check the fit before cutting your actual material. If it is too tight or too loose, adjust the jig accordingly.

◉ PREPARATION

Eliminate play with the movable stopper. When moving the base plate back and forth, ensure there is zero play by adjusting the movable stopper of the jig (see page 148 for the 'Box Joint Jig Structural Diagram').

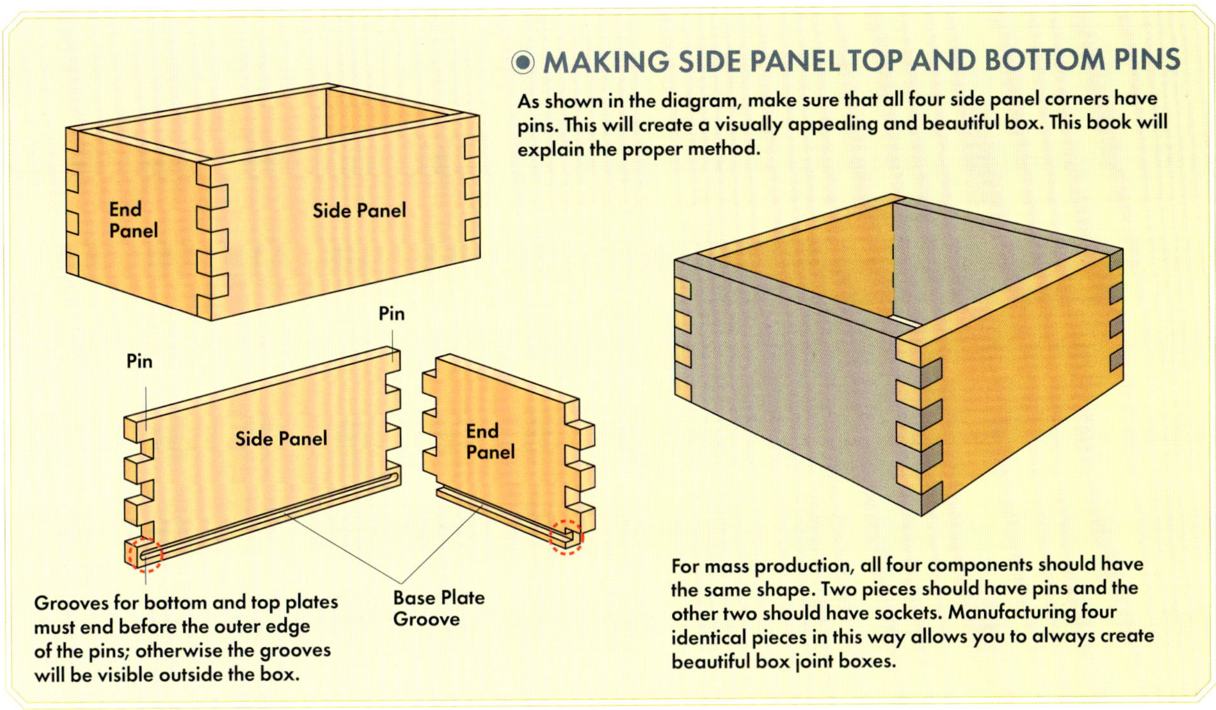

◉ MAKING SIDE PANEL TOP AND BOTTOM PINS

As shown in the diagram, make sure that all four side panel corners have pins. This will create a visually appealing and beautiful box. This book will explain the proper method.

For mass production, all four components should have the same shape. Two pieces should have pins and the other two should have sockets. Manufacturing four identical pieces in this way allows you to always create beautiful box joint boxes.

ADJUSTING DISTANCE BETWEEN SPIRAL BIT AND KEY MATERIAL

1

As shown, set the distance between the spiral/straight bit and key material to the bit's diameter. In this case, we used a down-cut spiral bit with a cutting and shaft diameter of ¼in (6.35mm). Loosen the adjustable fence screws and use a bit of the same size as the spacer between bit and key. Make this adjustment while slowly rotating trimmer bit by hand. Once distance is set, tighten the screws.
IMPORTANT Before processing actual material, make a test cut on scrap. Adjust the fit as needed.

SETTING SPIRAL BIT HEIGHT

2

Set the bit to match the material thickness. If you are using a hose clamp stopper on your trimmer, set it to this depth (see page 57 'Height Gauge/Depth Gauge').

STEPS FOR CREATING BOX JOINT PINS AND SOCKETS

3 Box joint sockets and pins should begin at the material top. This ensures that the top pin width aligns with the key. You'll need to turn the material over to process both ends, so follow the steps carefully for the best results.

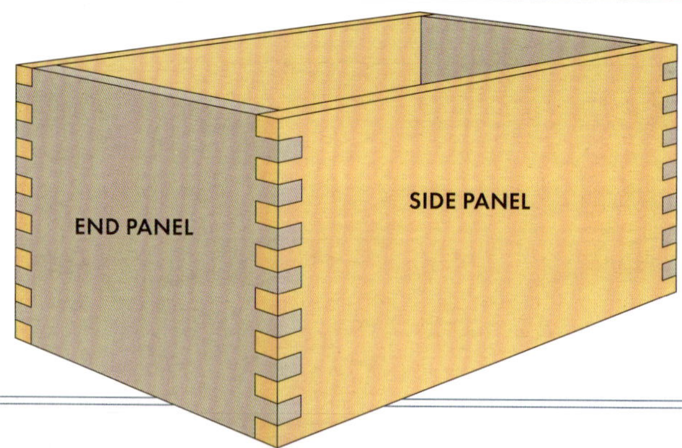

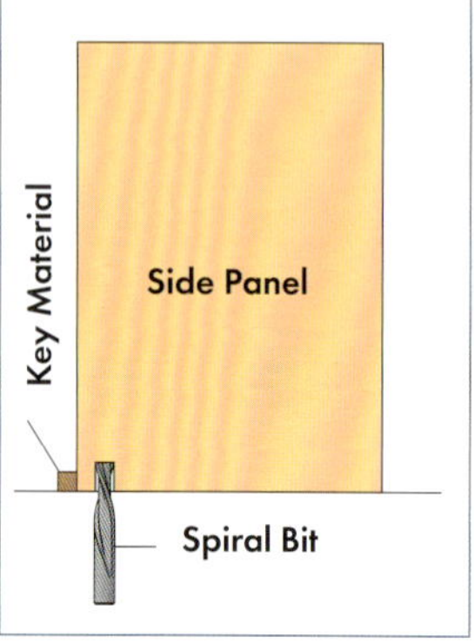

A Align the top edge of the side panel with the key and cut the initial socket.

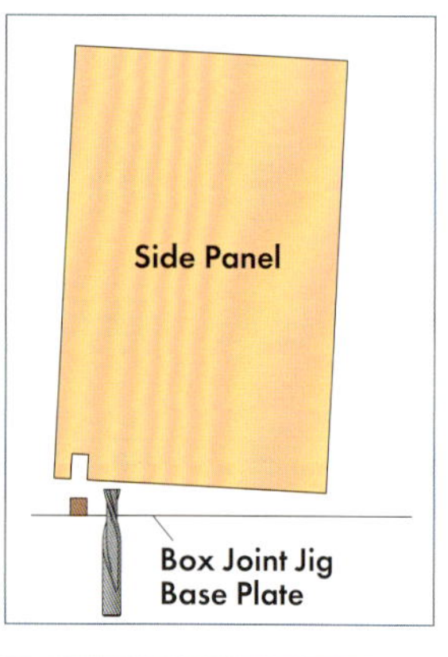

B Insert the key into the initial socket cut.

C Cut the second socket.

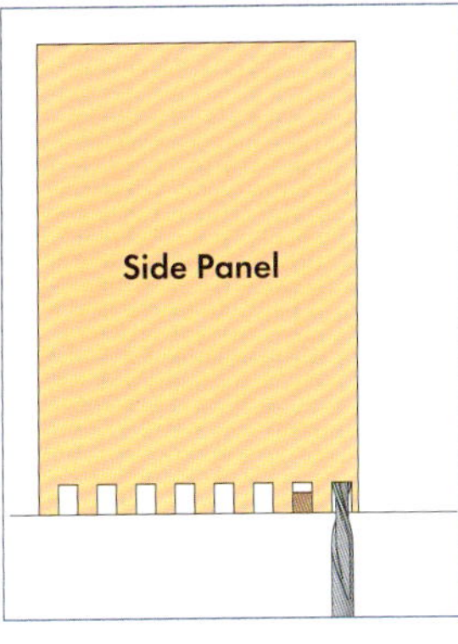

D

Proceed all the way to the end.

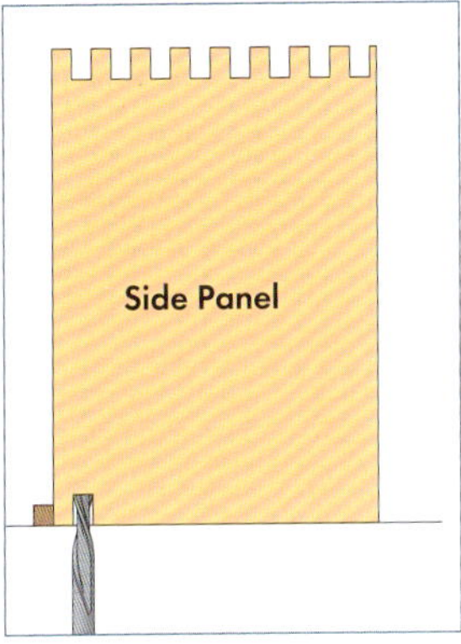

E

Flip the material vertically and align with the key material to machine the first socket in the same way as section A. When flipping the material, maintain its orientation.

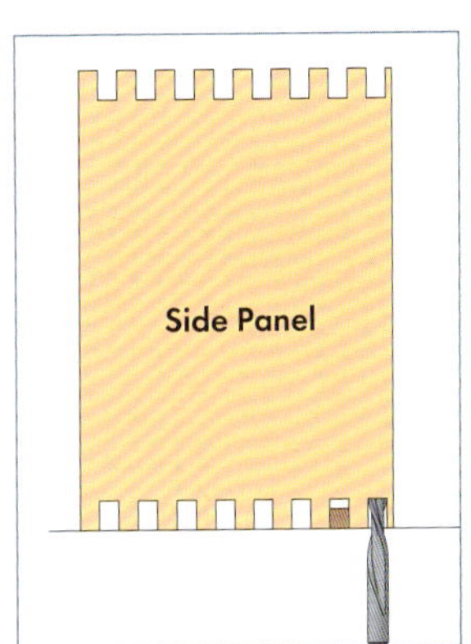

F

Cut the sockets until the end of the material.

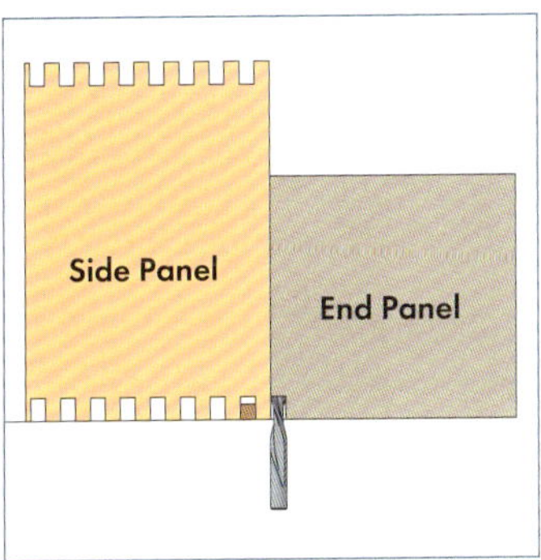

G

Use the pin at the top edge of the side panel (see diagram) for initial alignment of the end panel. Cut the end panel's first socket in this position.

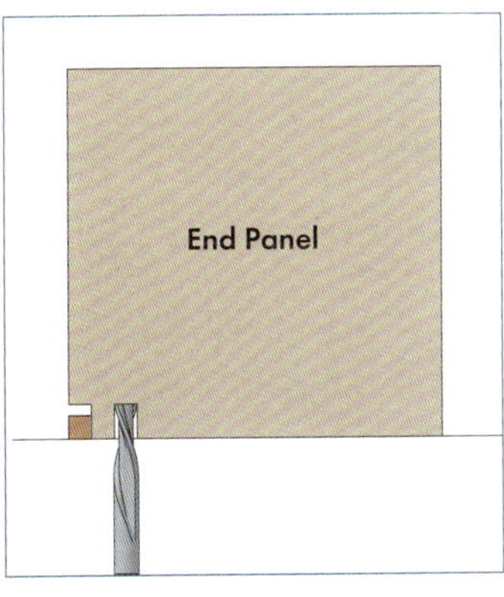

H

Align the first socket using the key and then cut the second socket.

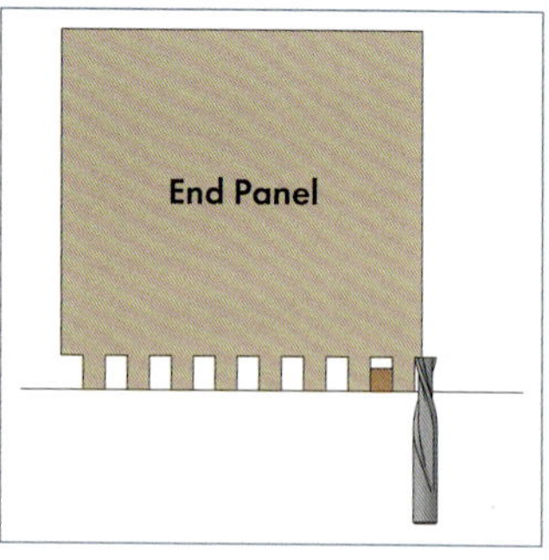

I

Continue step-by-step processing until you reach the end.

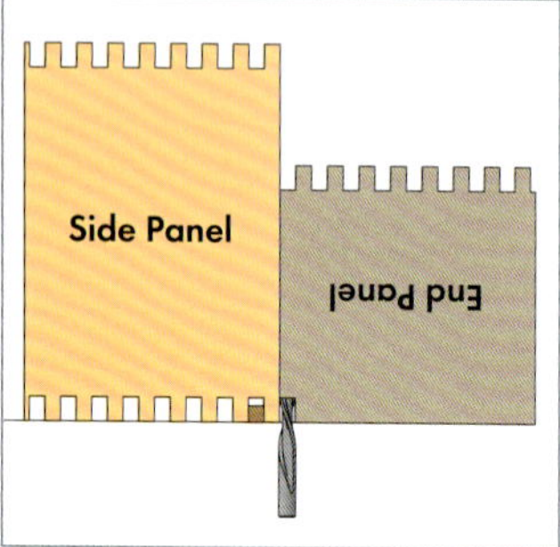

J

Flip the short side panel vertically and use the side panel for the initial alignment. Machine the first socket.

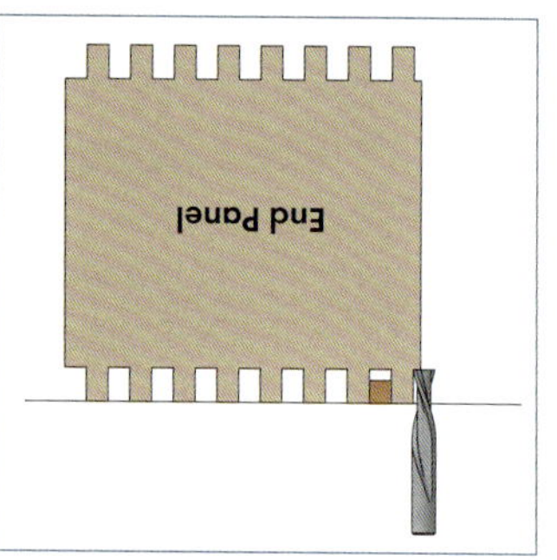

K

Continue step-by-step processing until you reach the end.

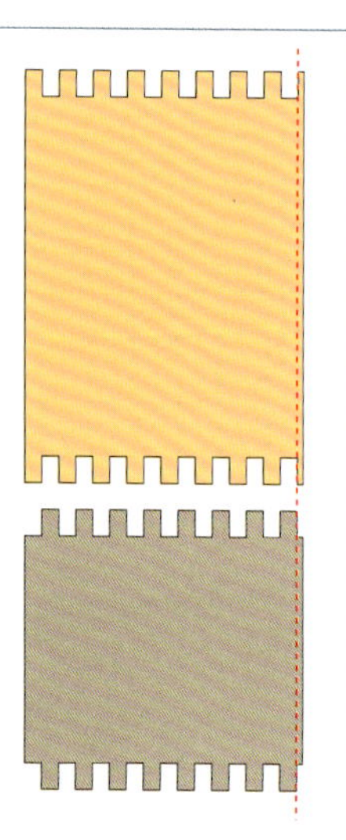

L

Trim partial-width pins and sockets at the bottom of the material and cleanly even out the height.

◉ INCREASE EFFICIENCY BY PROCESSING TWO BOARDS AT ONCE

For efficiency and improved precision, you can stack two side panels or two short end panels together and process them. Place outer faces so they touch each other. This prevents chipping by the trimmer bit. To secure the two boards, you can either use clamps or temporarily secure them with double-sided tape cut into four small pieces about ⅜in (10mm) square. Be careful not to use pieces of tape that are too large, as it will be difficult to separate the boards after processing. The processing procedure is the same as when machining a single board (see page 172).

FINE ADJUSTMENT/METHOD FOR SHIFTING MOVABLE FENCE LEFT AND RIGHT

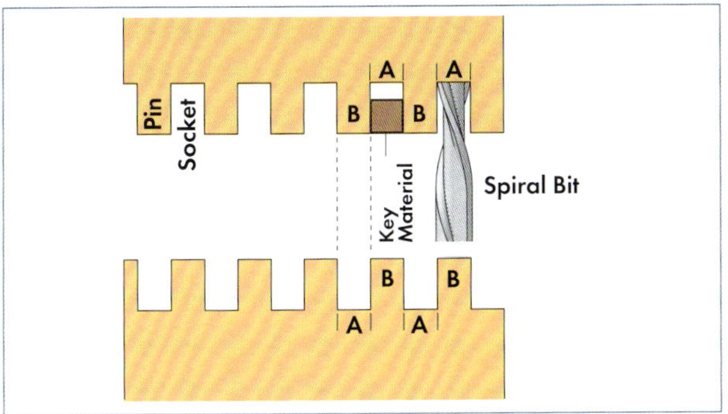

4 The socket width is determined by the spiral bit's cutting diameter. This width cannot be changed. Pins are created by the distance between the spiral bit and key. Its width can be adjusted by moving the movable fence left or right.

5 Fit is adjusted by changing either the pin width. If it's too tight, narrow the pin. If it's too loose, widen the pin. Measure the width of the pin and socket with callipers. Calculate the difference, and adjust using a paper shim (spacer).

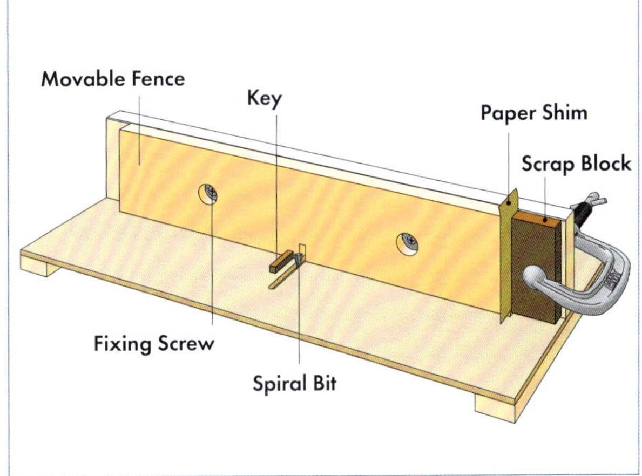

6 Fine adjustments to the movable fence can be made by shifting it left or right. Use paper shims and scrap wood blocks to account for differences between sockets and pins. Paper shims act as spacers, allowing for tiny changes to the movable fence. Moving the fence left increases the distance between the key and spiral bit, thus widening box joint pins. This method is used when the fit is too loose:

1. Place the scrap wood block against the movable fence and clamp in place.
2. Loosen the movable fence screws and insert a paper shim between the fence and scrap wood.
3. Push the movable fence against the scrap wood block and tighten the screws.

This shifts the movable fence left by the paper shim's thickness. Moving it to the right brings the key closer to the spiral bit, narrowing the box pins. If the fit is too tight, this method can be used to adjust it:

1. Place a paper shim between the movable fence and the scrap wood block and clamp in place.
2. Loosen the screws securing the movable fence and remove the paper shim.
3. Push the movable fence against the scrap wood block and tighten the screws.

This shifts the movable fence right by the paper shim's thickness.

BASE PLATE GROOVE

Cut base plate grooves into four boards after the pins and sockets are finished. We need to limit the pin board groove length to ensure the grooves are hidden when assembled (see page 155). To do this, attach two stoppers to the fence (circled in the photo). In this case, craft sticks were simply attached with double-sided tape. Set the groove depth to ¼in (5mm) with a height gauge. Also, adjust the trimmer's stopper (hose clamp). A ⅛in (3.2mm) spiral bit is used here (the base plate is $\frac{5}{32}$in/4mm thick so, after the first groove cut, move the fence back the necessary amount and perform a second cut to widen the groove).

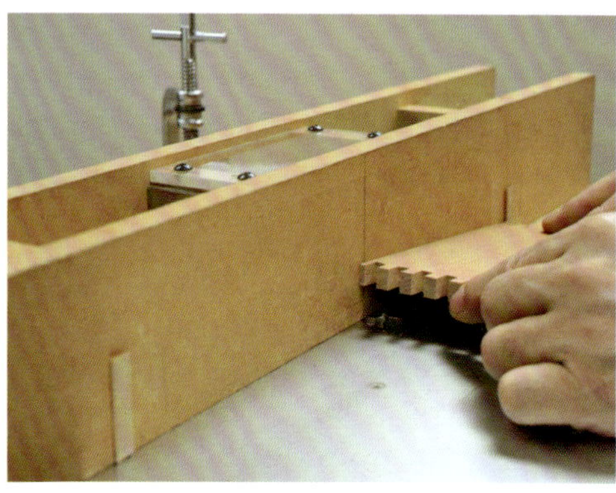

8 Place the material against the right side stopper and lift the left side to avoid contact with trimmer bit. Lower carefully to begin the groove cut.

9 Slowly move the material to the left.

10 Once the material touches the left stopper, lift the right end to remove it from the trimmer bit.

11 If the groove enters a socket, press the material against the fence and move it from right to left. Then move the fence back to widen the groove.

Grooving is now complete. Cut off any excess created by box joint cutting, to match up the pin sizes (see page 158, step L).

BASE PLATE

13 Dry fit and measure the inner dimensions. The groove depth is ¼in (5mm) on both sides. Add ½in (10mm) to the inner dimensions. Cut the bottom plate to 3⁄64in (1mm) less than that value. This will ensure that the base plate doesn't become too large. Before gluing, smooth the inner surface with a plane or similar tool.

14 Apply glue to the box joint areas and insert the base plate to fix it in place. (Note: The bottom plate is not glued).

15 Clamp securely. Use a square to check for right angles inside the box.

16 Sand the bottom on a flat surface to correct any height discrepancies.

17 Use a plane to correct upper discrepancies, which require a smoother finish.

18 Finish the side surfaces with a plane. Steps 17 and 18 can also be done with sandpaper.

Box Joint Inro Box

I will now introduce a uniquely designed box where the lid and body are fitted together using box joints instead of the more common inro style.

After cutting one long piece vertically to divide it into the lid and the body, cut box joints in the edges (bottom left photo). Fit the two pieces together and temporarily hold them with masking tape, treating them as a single piece. Then, cut them into the side and end panels that will form the box.

This beautiful box can be made by cutting the material at the boundary between the pins and sockets of the box joint (see diagram on page 165). Make sure to factor in the fact that you will be cutting between pins and sockets when deciding on the size of the box.

Also, note that it is necessary to make the fit a bit loose. If it's too tight, it will be difficult to open and close the lid.

Box joint processing is performed using a trimmer table and box joint jig. For detailed explanations, see 'Constructing a Box Joint Jig' on page 146 and 'Box Joint Box' on page 154. Additionally, you can make your own 45-degree inclined board for mitre cutting (see page 176). Mitre cutting can also be done with a mitre trimming platform using a hand plane.

Fit together the two box joint cut pieces. Then, cut the side and end panels.

A 45-degree inclined board for cutting 45-degree mitres. Used with a trimmer table.

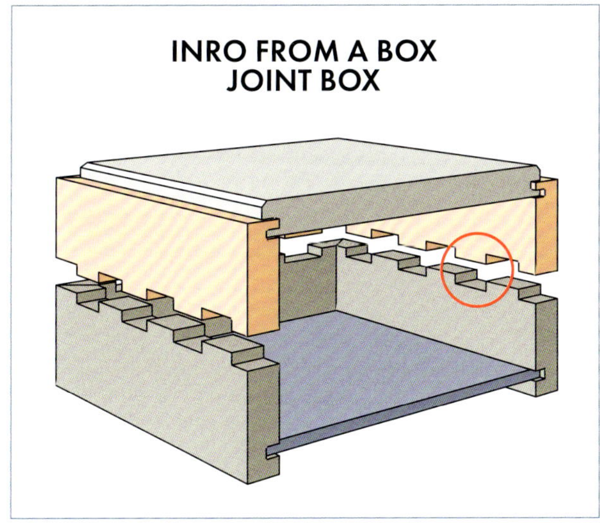
INRO FROM A BOX JOINT BOX

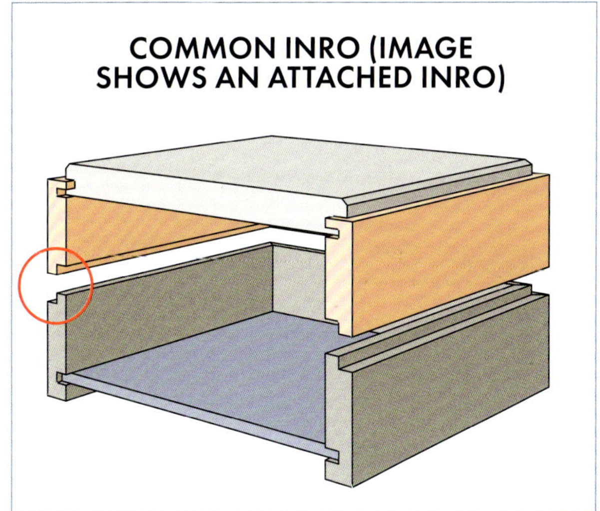
COMMON INRO (IMAGE SHOWS AN ATTACHED INRO)

PROCESSING BOX JOINTS

The trimmer bit has a cutting diameter of ¾in (19.05mm) and a cutting length of ½in (12.7mm). The cutting depth is approximately ⅛in (3mm), so the key's thickness will always be less than that. Each of the two pieces that will become the lid and body need to be rabbeted in the same manner, but they must be offset by the router bit diameter. Cut the first pin on each board separately and then match them up later. Similarly, process the lid material by first cutting sockets on each board. To prevent chipping, it's best to use a waste board. This waste board does not need to be long; it can be reused. The light-coloured plywood is the waste board.

2

After cutting two pins in the lid material, process the body. Setting up the body requires the lid material. Fit the already processed lid part into the key material as shown. Then, butt the body material up to it. Now, we can see the position for processing the first socket. Clamp on a waste board and cut the first socket. Next, process the second socket using the key as a guide. We are ready to process both materials together in an offset state.

3

We will process the entire length of both materials as a single piece. The photo shows the final section being cut. The material is quite long and unwieldy, even when held by hand, so it's clamped to the jig.

4

Check the fit, and if there are no issues, interlock the teeth and secure in position. You can use masking tape or similar materials for this.

◉ CUTTING SIDE PANELS AND END PANELS

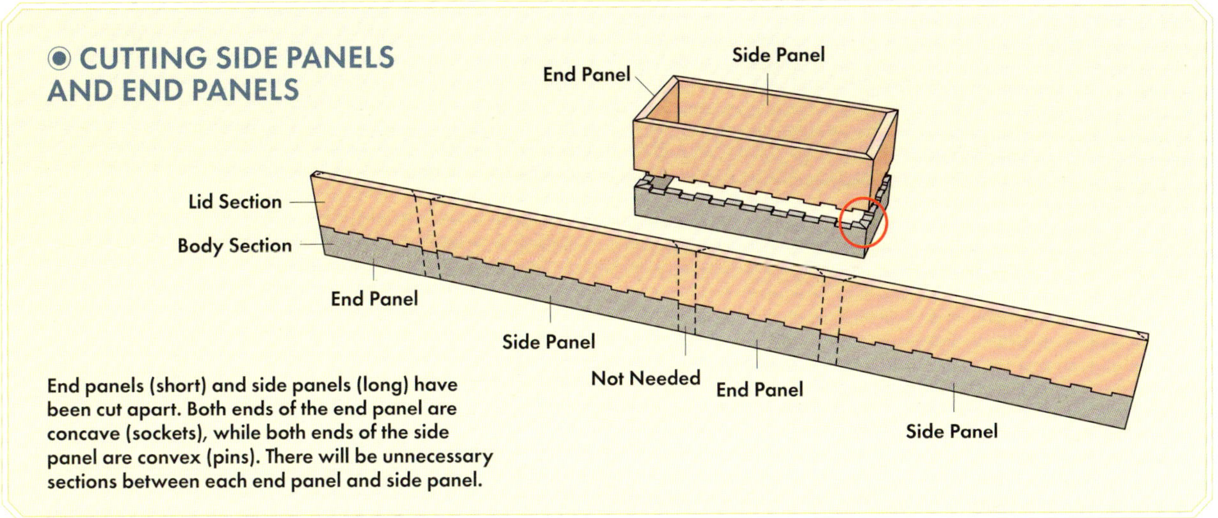

End panels (short) and side panels (long) have been cut apart. Both ends of the end panel are concave (sockets), while both ends of the side panel are convex (pins). There will be unnecessary sections between each end panel and side panel.

MITRE PROCESSING

5 Use a 45-degree inclined board (see page 176) to create the mitre joints on the ends of the cut pieces.

6 Clamp two pieces to the jig and process as one. The bit height determines the amount of material removed. Intentionally climb cutting helps reduce chipping. Move the jig so that the trimmer bit rotates anticlockwise around the material, creating a climb cut. Once one side has been finished with the 45-degree cut, mark the finishing position for the 45-degree cut on the opposite side.

7 For the 45-degree cuts on the opposite side, process each piece individually to determine the material length. Continue cutting until the finishing position is reached.

8 Use rubber bands, or something similar, to temporarily assemble the pieces and check for gaps.

9 Measure the internal box dimensions and cut the base plate and top plate to 11⁄32in (9mm) larger than that measurement. This accounts for the groove depth of 1⁄4in (5mm) on each side, totalling 3⁄8in (10mm), but subtracting 3⁄64in (1mm).

GROOVE CUTTING THE BASE PLATE

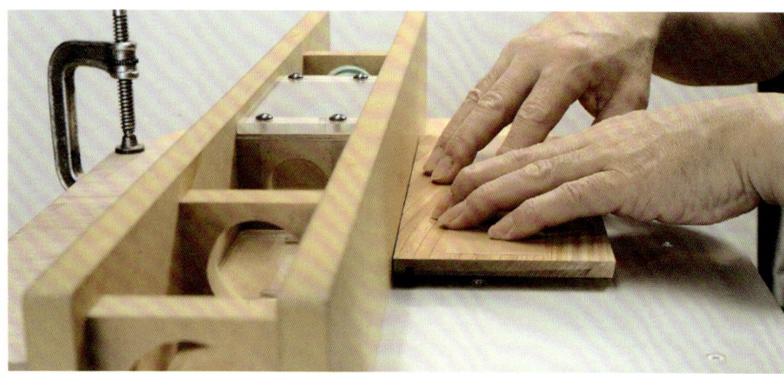

10 A 1⁄8in (3.2mm) down-cut spiral bit is used for processing the base plate groove. The groove depth is 1⁄4in (5mm). The depth is gradually achieved over several passes. It's necessary to widen the grooves because the base plate thickness is only 5⁄32in (4mm). For a detailed explanation, see page 83.

11 All materials have been grooved. The down-cut spiral bit minimizes splintering on both sides of the groove, resulting in a clean finish.

JOINT PRODUCTION FOR THE TOP PLATE AND LONG BOARDS

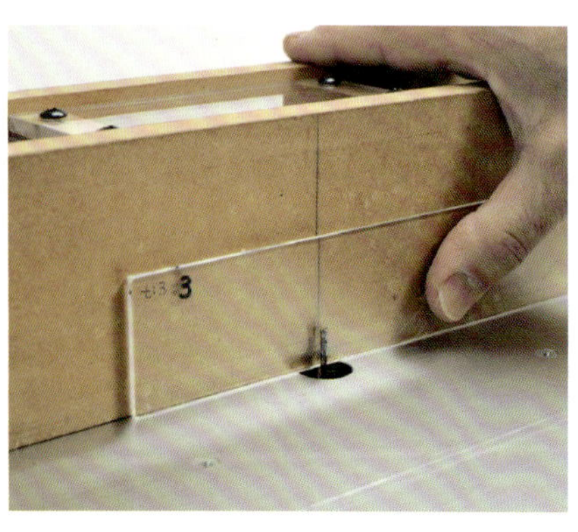

12 The bit used for this particular groove processing is also a 1⁄8in (3.2mm) down-cut spiral bit. Extend the bit to its maximum and place a 1⁄8in (3mm)-thick acrylic plate between it and the fence. This accurately sets the space between the bit and the fence to 1⁄8in (3mm). The spiral bit has a helical blade, so it will always make contact with the acrylic plate at its outer circumference. This makes it easier to set the bit and fence spacing between compared to a two-flute straight bit, i.e. since it is equivalent to a 1⁄8in (3mm) diameter round bar, you can just place the acrylic plate between the bit and the fence to set the spacing to 1⁄8in (3mm).

◉ SMALL TENON HALF-BLIND RABBETED DADO JOINT

The end panels and side panels will be joined to the top plate with a modified version of the small tenon half-blind rabbeted dado joint (abbreviated as 'drawer joint') as shown in the diagram. Although complicated at first glance, this joint is actually quite convenient as both parts can be processed with just one set distance between the spiral bit and fence. This distance is the same as the diameter of the spiral bit, but it should be set slightly smaller just to be safe. The depth of the groove will be just over ¼in (5mm). See page 196 for the procedure.

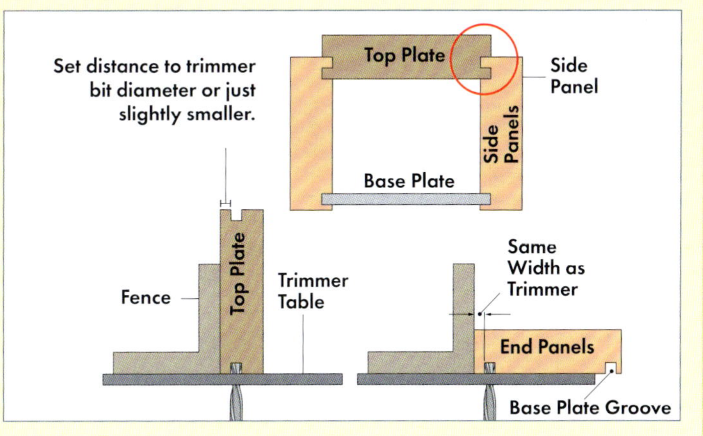

13 The top board is cut beginning with the end grain. The order is end grain → edge grain → end grain → edge grain. The inside of the top plate is placed against the fence.

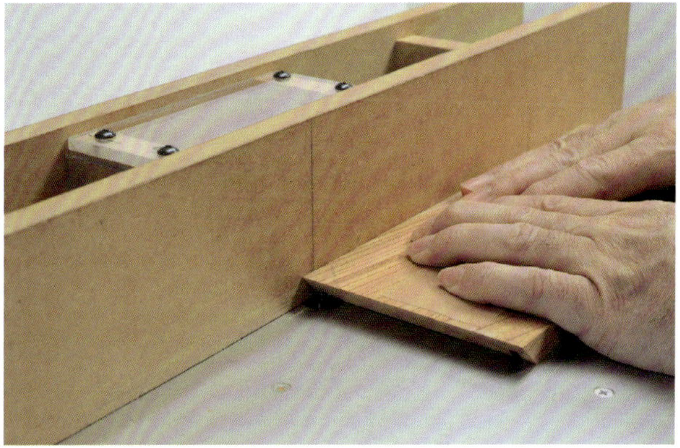

14 The end panel and side panel will be processed with the same groove setting. Depth will be achieved gradually over several passes.

15 The grooving is complete.

DRY FITTING AND ADJUSTING

Perform a dry fit and check for any issues.

17 The fit between the side panel and top plate was slightly tight. I corrected it with a chisel.

GLUING

Remove the masking tape used for temporarily holding lid and body together. Assemble the top and bottom sections of the box. First, connect the side and end panels with masking tape. Be sure to pull the tape tight to avoid any gaps.

19 Apply glue to the joints and assemble. Wrap the end and side panels around the top plate. Do not apply adhesive to the top plate! After wrapping, add masking tape to the last corner.

Place cellophane tape on the corners to prevent excess adhesive from accidentally bonding the top plate and body together. After this, use belt clamps to secure the top plate and body.

21 Tighten the three belt clamps. Place aluminium angle guide at each corner to prevent misalignment and to avoid crushing the corners with belt clamps. Wipe off any excess adhesive and check if everything is square. If adjustments are needed, use a string tied to the diagonally opposite aluminium angle guide to correct.

FINISHING

22 Use a plane to smooth any bulging at the joint between lid and body. Then, apply a bevel. Here, a Western plane is being used. You can also finish with sandpaper.

BEVELLING

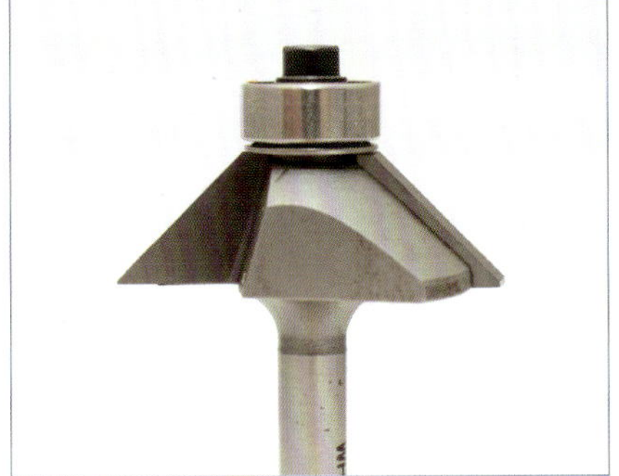

23 To make sure that the box joint has clearance, we will bevel the body edges. This will also enhance the three-dimensional effect when the top plate is attached. The bit used for this is a bearing-guided bevel bit. Start bevelling away from box corners to avoid cutting too deeply with the trimmer bit. This can be quite dangerous!

24 The saw-toothed, box joint edges are now clearly visible, highlighting the unique feature of this box.

Paper Towel Holder

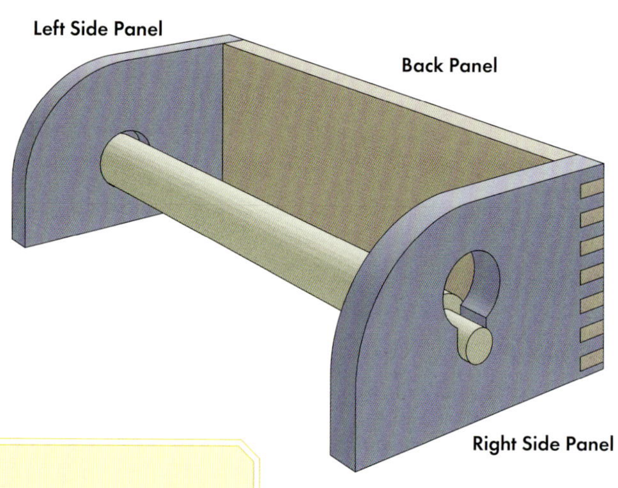

For this project you will learn how to thin dowel ends and create shoulders. Additionally, a method for replicating shapes using a template and a flush trim bit will be introduced. For instructions on making the template, see page 175. Please read 'Processing Curves Using a Template'.

TOOLS TO BE USED ARE AS FOLLOWS:

- * Drill Bits: 15mm, 30mm (32mm = 1¼in is also ok)
- * Forstner Bit: 32mm (= 1¼in)
- * Trimmer Bit: Spiral Bit (6.35mm = ¼in is also ok)
- * Flush Trim Bit (Cutting Length: 9/16in/14mm or more)
- * DIY Box Joint Jig (pages 146–153)
- * Trimmer Table

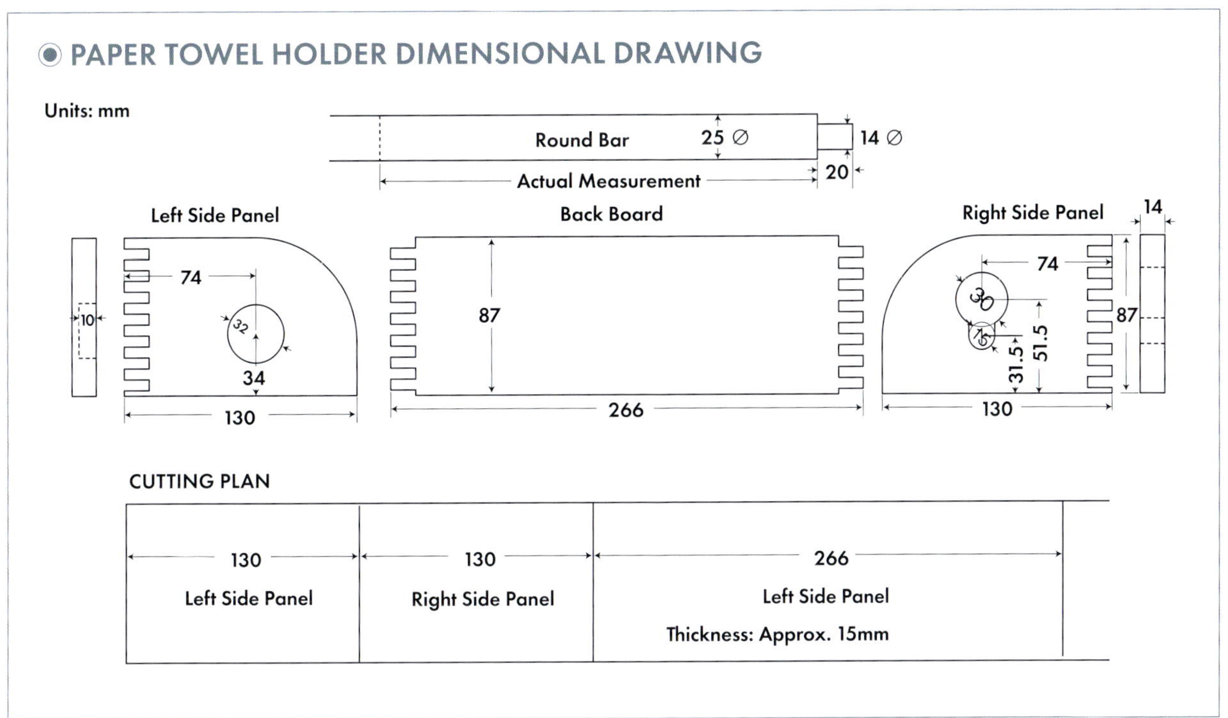

CREATING SHOULDER ON ROUND BAR

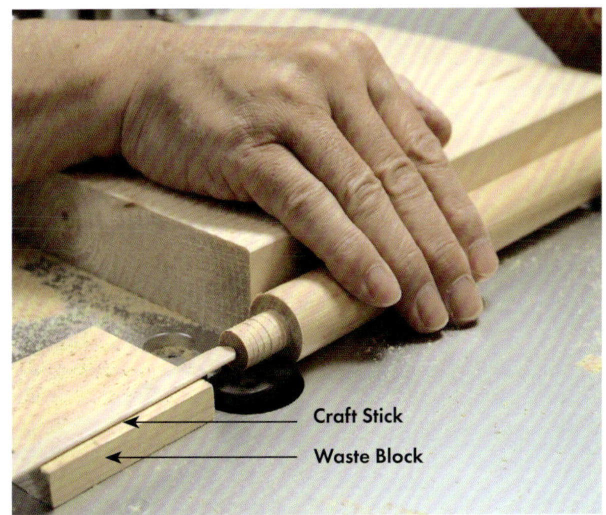

1

Create shouldered dowels using a trimmer table. This shoulder acts as a stopper to prevent the dowel from slipping out of its holder. Clamp thick board to the trimmer table as a fence, and, while holding the dowel against it, rotate the dowel by hand. The direction of the rotation follows the arrows. The length of section to be thinned is determined by the craft stick stopper position. The stopper's height is around the dowel's centre. Both the stopper and scrap block underneath the table are fixed with double-sided tape. The thickness of the part being processed depends on the trimmer bit height. I used a dowel with a 1in (25mm) diameter, and the processed section was reduced to 9/16in (14mm) to fit a hole that is 11/16in (15mm) in diameter. A large-diameter, short-length straight trimmer bit was used. The bit sizes were: 1¼in (32mm) and ¾in (19mm). Begin the process at the dowel's tip, gradually rotating and moving towards the craft stick stopper. Adjust the trimmer bit height gradually over several passes to achieve the desired depth.

BOX JOINTS

2 Simultaneously perform box joint cutting on both side panels. Attach two panels together using four pieces of double-sided tape, each cut to about ¼×¼in (5×5mm). Place the outer sides of panels back to back. Single clamp for stability. Use small pieces of double-sided tape to ensure that the two boards can be easily separated after box joint cutting.

3 Separate the side panels before proceeding with this task. Use one processed side panel to align the back panel for cutting (see page 156). Cut the first socket.

4 Continue sequential cutting to complete the box joint grooves on the back panel.

5 Dry fit the pieces, and use sandpaper to remove any burrs.

OPENING HOLES IN SIDE PANELS

6 Drill a ⁹⁄₁₆in (15mm) diameter hole into the right side panel as per dimensions. Set up a fence on your drill press and drill with the material held against that fence.

7 Move the material along the fence, drilling holes to create an oval through-hole.

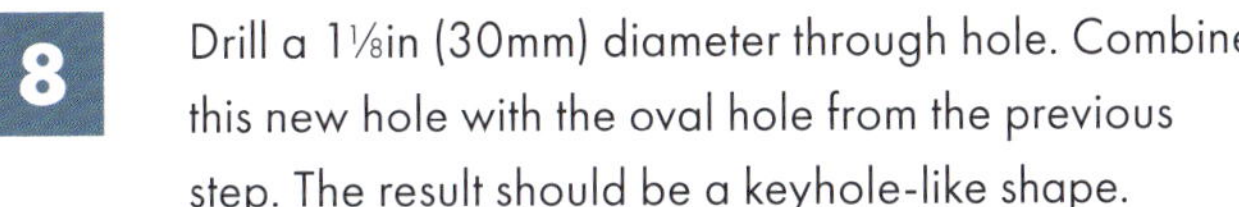

8 Drill a 1⅛in (30mm) diameter through hole. Combine this new hole with the oval hole from the previous step. The result should be a keyhole-like shape.

9 Drill a 1¼in (32mm) diameter hole in the left side board using a Forstner bit, ⅜in (10mm) in depth. Do NOT drill all the way through.

PROCESSING WITH A TEMPLATE

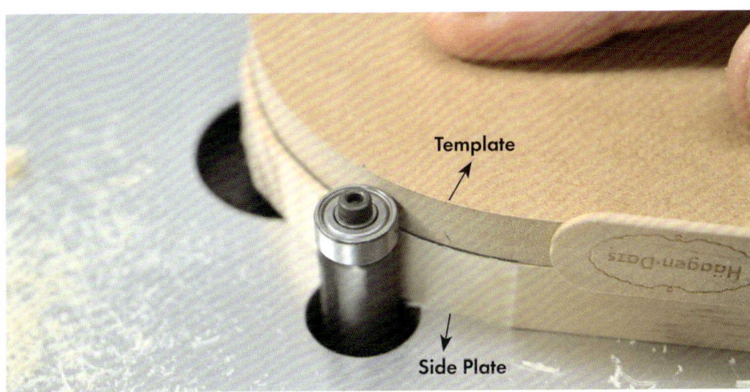

10 Attach a template made from MDF to the side board using double-sided tape. Cut along the template using a flush trim bit (with bearing) to significantly round the side board corners. For details on the template, see page 175.

CHAMFERING

11 Bevel the edges around the hole. Adjust the chamfer bit height and make a test cut on scrap before proceeding.

PLANING THE INSIDE FACE

12

It's difficult to smooth the inside surfaces after assembly. Now that the parts are finished, sand or plane the inside surface. Here, a Western-style jack plane was used.

ASSEMBLY AND GLUING

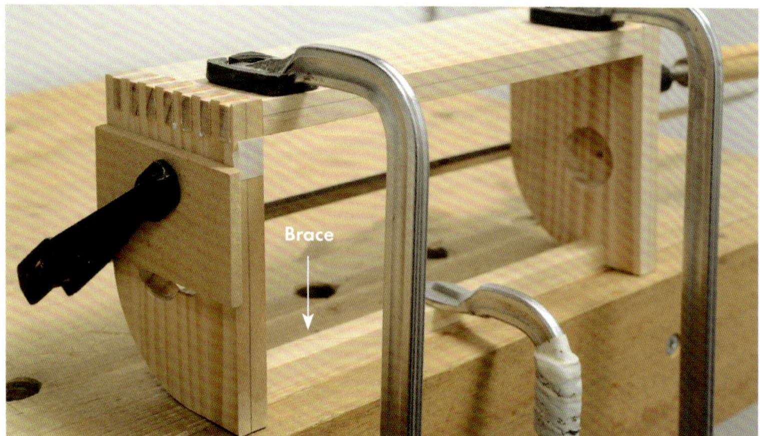

13

If there are no issues with dry fitting, apply glue and assemble. Secure the top and bottom with two clamps. Add one long clamp to tighten the sides. Here, a temporary brace is used to maintain the overall shape.

CUTTING ROUND ROD

14

Cut the rod prepared in step 1 to length. The photo shows the round rod itself being used to mark the correct distance. Cut the rod based on this distance. See the dimensional drawing on page 171; it's the part labelled 'Actual Measurement'.

PLANING

15

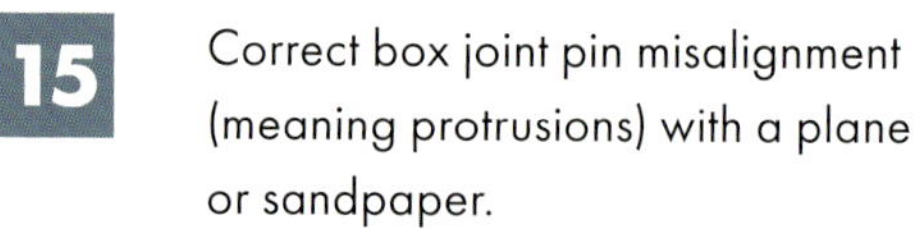

Correct box joint pin misalignment (meaning protrusions) with a plane or sandpaper.

16

Correct the top and bottom edges of the back panel to complete the project. Here a block plane was used.

◉ PROCESSING CURVES USING A TEMPLATE

To create a large curve on the edge of the material, attach a template made from MDF with double-sided tape. Then, use a flush trim bit to shape the curve along the template. This makes it easier to create the same shape each time. The craft stick attached to the template acts as a temporary fence for positioning. Note: If the material is roughly cut along the curve with a saw in advance, the shaping process becomes easier (the left side plate in the photo).

How To Make a Template

Sand a roughly cut curve clean to make a template. The sanding block in the photo is something I use quite regularly. It's made by wrapping sandpaper around an aluminium square tube and attaching it with double-sided tape. If you place a thin board underneath, when you move the sanding block back and forth, the thin board will move along with it. Also, be sure to place the template on a scrap board, lifting it slightly. You'll get a nice clean curve, with perpendicular cross-sections, if you finish up to the pencil line. Use plywood or MDF, with a thickness of ⅜in (9mm) being suitable.

Set the height so that the flush trim bit bearing aligns with the template, and proceed with shaping. Move the material to avoid climb cuts. In the photo, the material is rotated anticlockwise during processing.

Regarding Grain Direction: Against the Grain and With the Grain

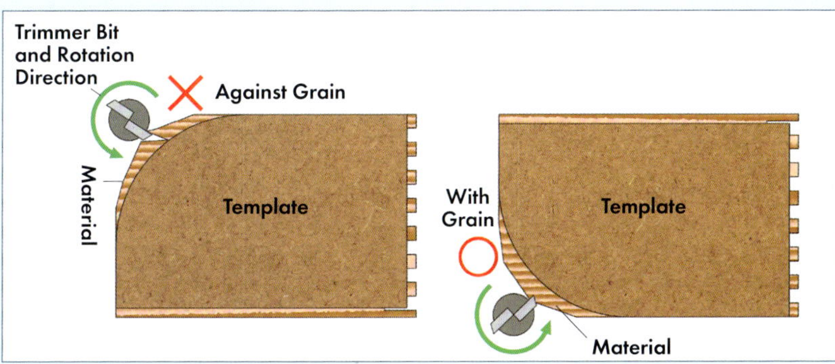

I've explained the cutting direction, but it's also important to cut along the grain. Orient the surface where template is attached as shown in the diagram, so you can cut with the grain. This reduces the likelihood of working against the grain.

Method for Producing a 45-degree Inclined Board

If you create a 45-degree inclined jig, it can be used for various purposes. This will expand the range of your projects (box joints, inro boxes, wall-mounted boxes, etc.). However, you may be hesitant as making a jig with an accurate angle can be difficult. Here, I'll introduce a method for making a jig that isn't all that complicated and ensures precise angles.

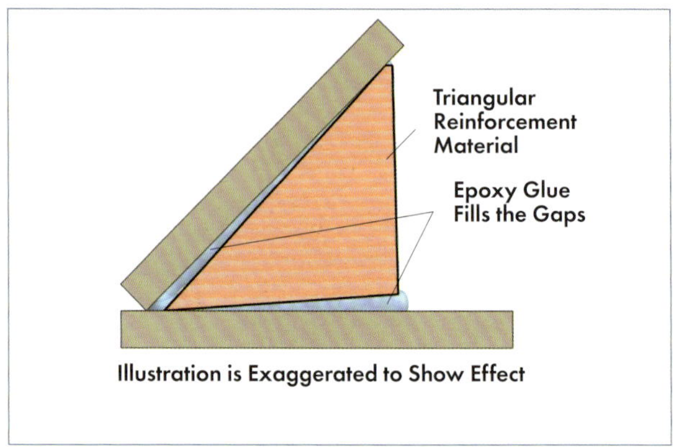

Illustration is Exaggerated to Show Effect

First, prepare two mitre gauges. The role of the double-sided, taped-down, mitre gauges shown in the photo is to hold two pieces of plywood at a precise 45-degree angle. Once the angle is set, add triangular reinforcements using epoxy glue. This glue also functions as a filler. Epoxy, which hardens when its two components (base and hardener) are mixed, does not shrink after hardening. By contrast, woodworking glue contains moisture and shrinks as it dries and hardens. Epoxy also has a high viscosity, similar to honey, and becomes even thicker during winter. This property is what we make use of here. You can create a highly precise jig by filling the gaps between the triangular reinforcement material and the plywood with epoxy glue. Use a thick board for triangular reinforcement material, as you will be applying generous amounts of epoxy adhesive to the cross-section.

CUTTING OUT TRIANGULAR REINFORCEMENT MATERIAL

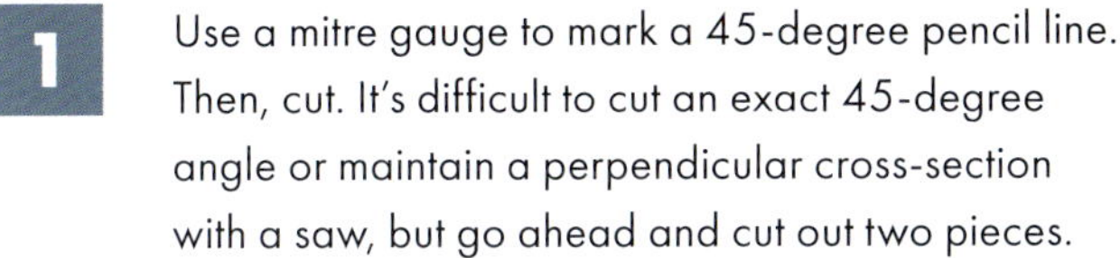

1 Use a mitre gauge to mark a 45-degree pencil line. Then, cut. It's difficult to cut an exact 45-degree angle or maintain a perpendicular cross-section with a saw, but go ahead and cut out two pieces.

2 The triangular reinforcement material is now cut out.

MARKING AND CUTTING THE INCLINED BOARD

3 Mark the positions of the triangular reinforcement material and mitre gauge with a pencil on ¼–⅜in (5.5–9mm) plywood. Then, divide the plywood into two pieces: the inclined board (left) and the base board (right). Sizing is up to you.

4 Apply double-sided tape to both sides of two mitre gauges, as shown in the photo. Here, removable double-sided tape is used. Also, attach a square rod with a width similar to the workbench with double-sided tape. This square rod will act as a fence.

5 Attach the thick part of the mitre gauge to the inclined board. At this time, align the tips of the inclined board and mitre gauge with the fence.

Attach the flipped-over inclined board while aligning the base board with the fence. Note that the inclined board should also be aligned with the fence. Both pieces of plywood should now be temporarily fixed at a precise 45-degree angle.

7

Temporarily fixed plywood pieces can easily come apart, so proceed to the next step without applying too much force.

8

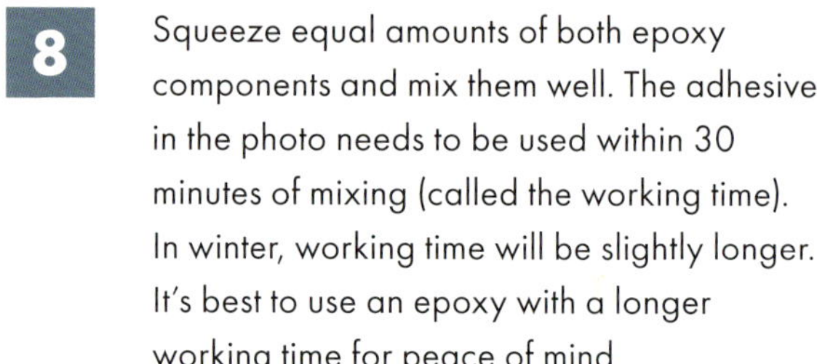

Squeeze equal amounts of both epoxy components and mix them well. The adhesive in the photo needs to be used within 30 minutes of mixing (called the working time). In winter, working time will be slightly longer. It's best to use an epoxy with a longer working time for peace of mind.

Apply a generous amount of epoxy adhesive to the bonding surfaces of triangular reinforcement material. Make it so that it overflows when attached to the plywood.

10 Align the reinforcement material with the pencil line and press into place. Do not press too hard, as this could dislodge the mitre gauge.

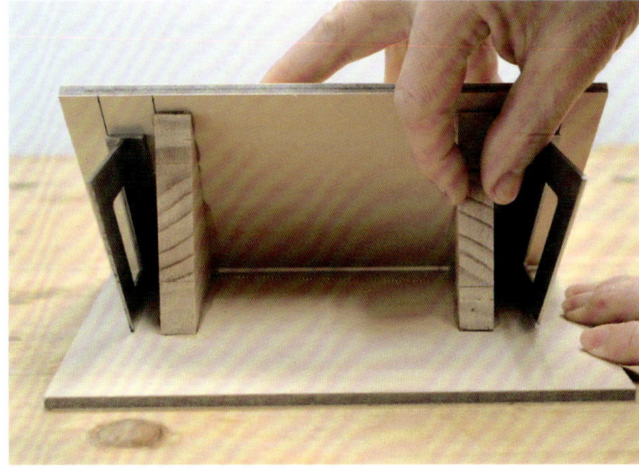

11 Confirm the position of both triangular reinforcement pieces.

12 Ensure that the adhesive overflows from the bonding surface. If there isn't enough adhesive, it won't bond properly, so apply more if necessary.

13 After allowing sufficient time for hardening, remove the mitre gauges to complete the process. Perform any additional work necessary for each jig.

◉ ATTACHING THE FENCE

Use a carpenter's square placed on a trimmer table when attaching the fence to the 45-degree inclined board. This will ensure that it's attached perpendicular to the table surface.

Desktop Chest

Next, we will learn about dado joints and drawer construction by producing a compact desktop chest of drawers. There are, of course, various methods for constructing drawers and here we will merely introduce one such method.

First, we will create the box that will house the drawers, and then we will make the drawers to fit the opening. We will use a saw for certain portions of the drawer-making process. To ensure straight, perpendicular, and precise cuts, it is recommended that we make a saw guide (see page 189). Feel free to make your preferred size guide, rather than strictly following our dimensions.

◉ DRAWER DIMENSIONAL DIAGRAM

Units: mm/t: thickness

Front Board (t:15)
Side Board (t:10)
50
60
190
15-20
250

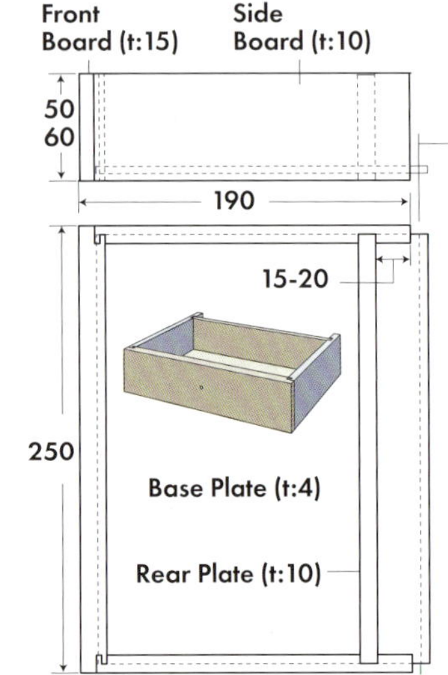

Base Plate (t:4)
Rear Plate (t:10)

Back Board

After drawers are complete, front board positioning occurs by adjusting the rear edge of base plate. Trim the base plate near the back board surface and adjust with a plane so that the front board and the front of the main box are flush.

External drawer dimensions are for reference only. In actuality, cut drawer components according to opening size.

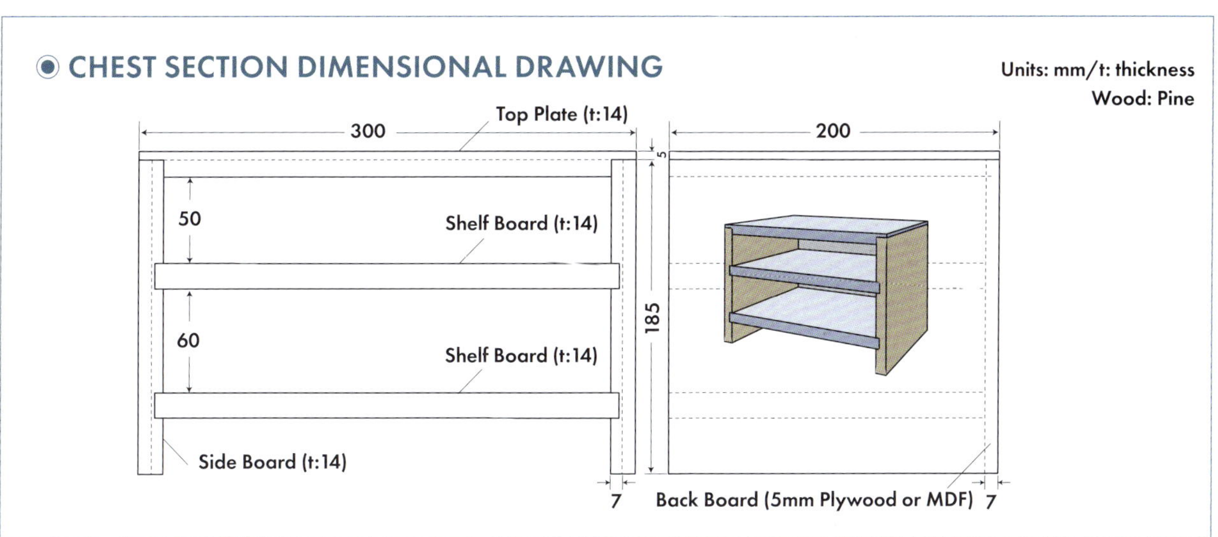

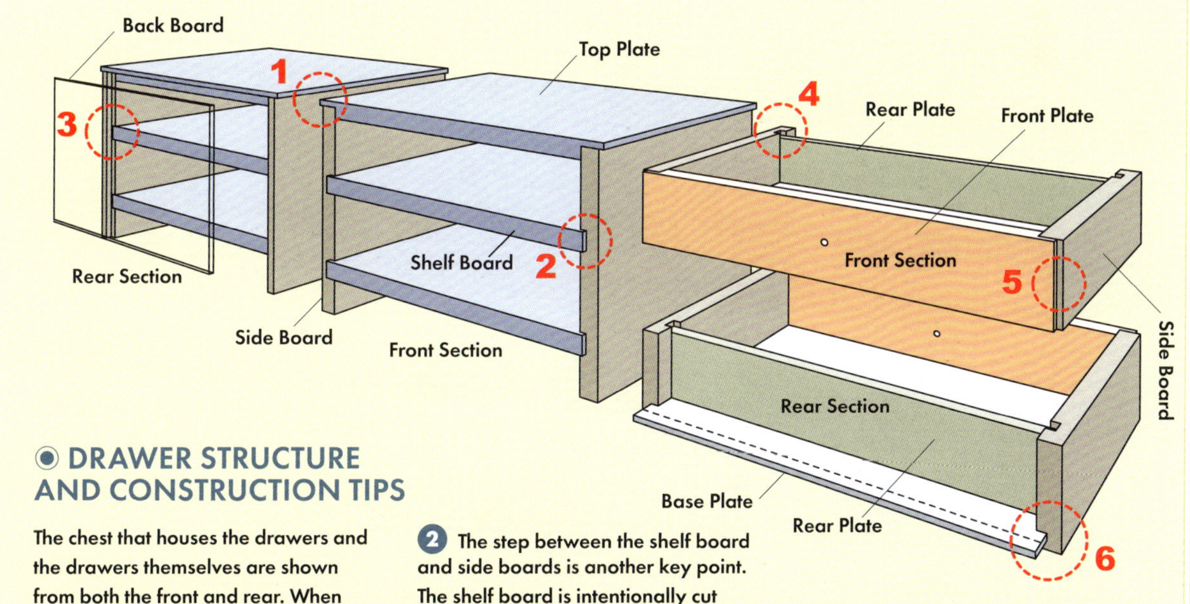

DRAWER STRUCTURE AND CONSTRUCTION TIPS

The chest that houses the drawers and the drawers themselves are shown from both the front and rear. When complete, the joints are flush and neatly finished. However, the joints are intentionally made with a slight offset, which will be planed down later to create the perfect finish.

1 In the half-blind rabbet joint between the top plate and side boards, the rabbet protrudes slightly like an overhang. After gluing, this overhang can be planed (or sanded) down, ensuring that the side boards are flush. If it were the opposite, it would be difficult to fix. Although exaggerated in the diagram, the actual protrusion is probably less than $^3/_{64}$in (1mm).

2 The step between the shelf board and side boards is another key point. The shelf board is intentionally cut to protrude slightly beyond the side boards and is trimmed afterward.

3 The depth of the shelf board stops at the rabbet where the back board will be inserted. The front of the shelf board is made to protrude slightly, and it is trimmed later.

4 Lowering the rear plate slightly below the top of the side boards helps mitigate snagging when drawers are pulled out.

5 The drawer's front board is cut to fit snugly. If the side boards behind it extend beyond the front board, the drawer won't fit properly. By making the side boards protrude just slightly beyond the front board and then trimming them to match, a perfect-fitting drawer can be achieved.

6 The base plate is cut slightly long and trimmed at the dotted line where it meets the back board. By adjusting the highlighted section, the position of the front board can be properly determined. The base plate is fixed to the front board with a single wood screw at the centre of its thickness. It's not glued, so removing the screw allows the bottom board to be pulled out for repairs or adjustments.

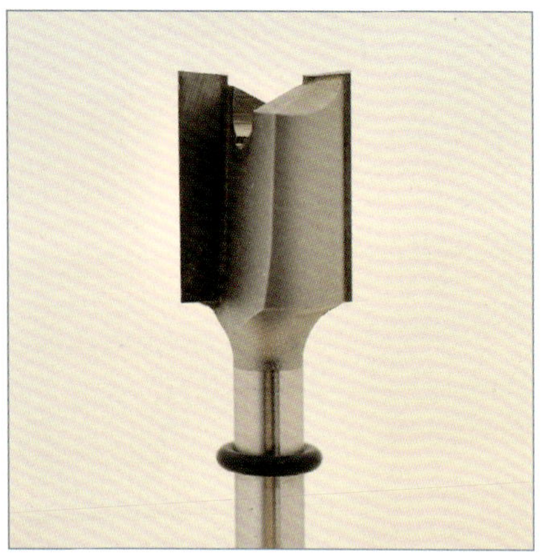

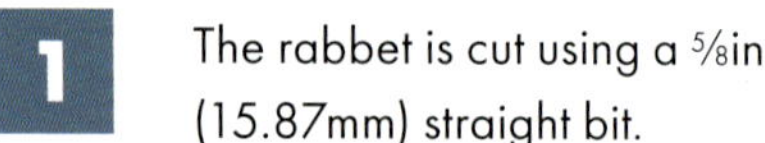

1 The rabbet is cut using a ⅝in (15.87mm) straight bit.

2 Cut out the top plate and two side boards based on the dimension diagram. Assemble temporarily. The shelf board lengths will be measured and cut after the rabbets on the top board are completed.

3 Adjust the distance from the fence to the outer edge of the bit to determine the rabbet depth.

4 Set the bit height to maximum cutting depth for the rabbet. Lock the trimmer lever at this height.

5 Gradually raise the bit, and cut the rabbet in several passes until the desired depth is reached.

6 To prevent chipping on ends, place the backing material behind the workpiece. This method is quite effective at preventing chipping. The offset area in the photo is the backing material. You can use scrap wood for this.

7 You can see the result of not using backing material in the photo at left. There's leftover material, and in some cases, chipping. The photo at right shows when backing material is used. The cut is clean and smooth.

8 Rabbet successfully cut to desired depth. Temporarily assemble and check the joints.

PROCESSING THE DADO AND SIDE BOARDS

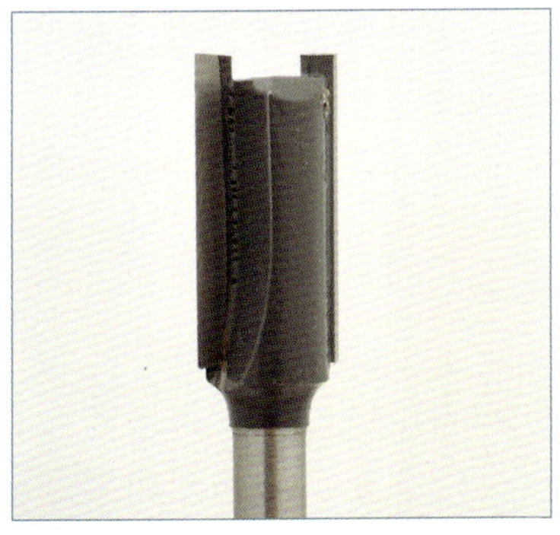

9 For this step, a ½in (12.7mm) straight bit is used. The groove width is 9/16in (14mm). Make a first pass with the smaller bit, then the second pass will widen the groove.

10 Pencil in a line to indicate the dado top. This will serve as the first pass marker. Use a square to align the cutting bit position. The side board is pressed against the fence at the top. Use a smaller bit hole in the trimmer base and position the fence on the side without a dust collection port.

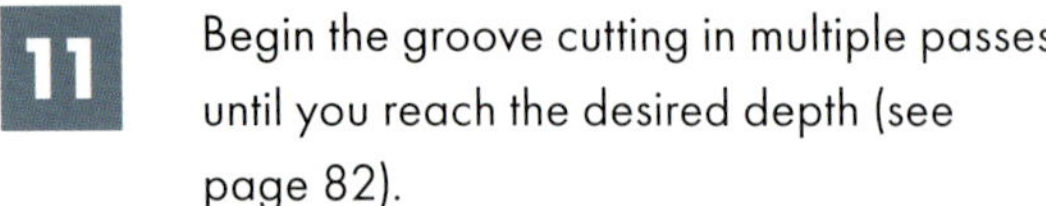

11 Begin the groove cutting in multiple passes until you reach the desired depth (see page 82).

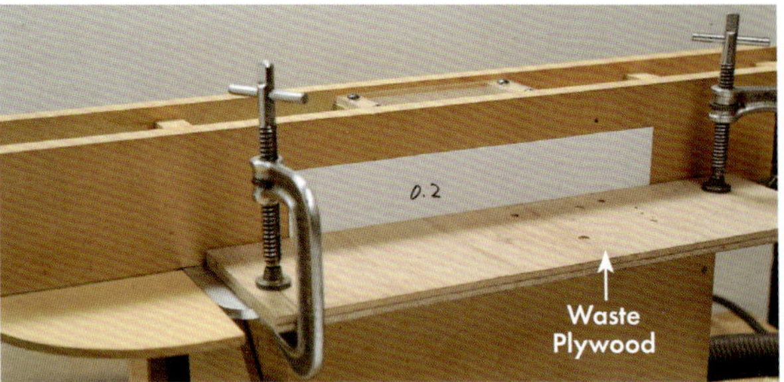

12 To widen the groove, measure the board thickness with callipers and subtract the first pass groove width. This value equals the distance to move the fence. Place a spacer behind the fence, held in place with scrap plywood. Clamp securely.

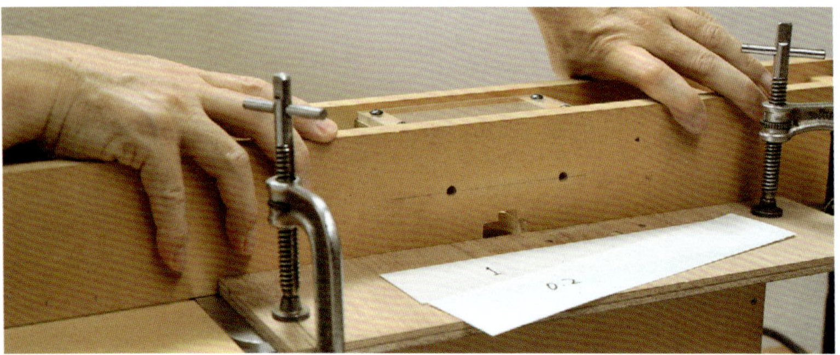

13 After clamping, remove the spacer and adjust the fence by sliding it back to the plywood. This moves the fence back. Make the second groove pass. If the joint feels tight during test fitting, insert a thinner spacer and adjust the fence slightly to widen the groove.

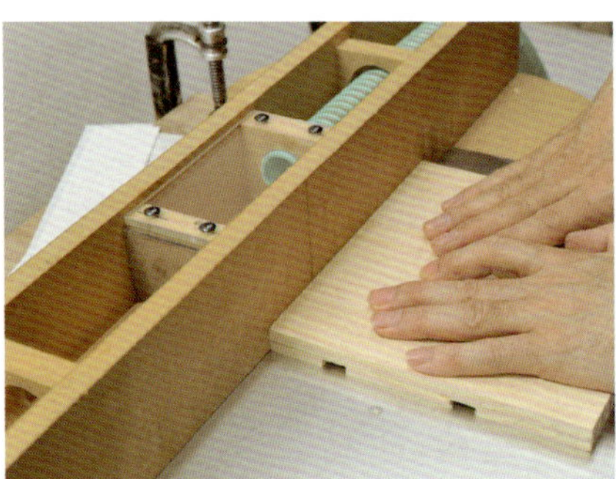

14 Repeat steps 9 through 13 to cut the second dado.

CUTTING REAR PLATE RABBET

15 Dry fit the parts, then cut a rabbet to fit the rear plate. For a rear plate of 1/4in (6mm), the rabbet should be approximately 9/32×9/32in (7×7mm).

16 Set the bit height to around 9/32in (7mm). Secure the hose clamp stopper on the trimmer. Perform cuts in several passes (see page 57).

17

The side panel and top plate rabbets are complete.

CUTTING SHELF BOARDS

18 To determine the shelf length, place the top plate on the front edge of the side panels and lightly clamp. Then, measure the distance from outside edge to outside edge of each rabbet. This value will be the shelf length.

19 The shelf must be flush with the back panel rabbet and should be cut so that the front edge protrudes slightly beyond the shelf.

GLUING

20 Apply glue to each part and tighten with belt clamp and F-type clamps. Check the corners are square.

21

On the left is a ¼in (6.35mm) diameter bit (for rear plate's mortise joint). On the right is a ⅛in (3.2mm) down-cut spiral bit (for front and side panel joints). The down-cut spiral bit minimizes splintering on both sides of the groove, resulting in a cleaner finish.

◉ DRAWER JOINT

This is a convenient method that allows you to process both front and side panels with just one fence setup. Since the front and side panels lock together structurally, it's ideal for drawer joints. See pages 196–197 for illustrations. Bits used are a ¼in (6.35mm) and a ⅛in (3.2mm) down-cut spiral bit. During fabrication, there will be a need to trim part of the processed material with a saw. See page 189 for 'Useful DIY Right-Angle Guide' and page 233 for 'About Aluminium Right-Angle Guide'.

22

Cut the front and both side panels so they fit perfectly in the opening. In the photo, the central panel is the drawer's front panel. Do not cut components to length yet. Keep them longer for now.

PROCESSING FRONT AND SIDE PANELS (DRAWER JOINT)

23

Assemble the drawer's front and side panels using drawer joints (see page 167, 'Small Tenon Half-Blind Rabbeted Dado Joint'). It's best to perform a test cut on the side and front panel scrap pieces before proceeding with the actual cuts.

PROCESSING THE SIDE PANEL GROOVES

24

A ⅛in (3.2mm) down-cut spiral bit is used. Place a spacer so the bit is positioned ⅛in (3.2mm) away from the fence. To ensure it's set in the correct position, the bit should be raised as high as possible.

25 Next, set the cutting depth. Estimate it to be about half the side panel thickness.

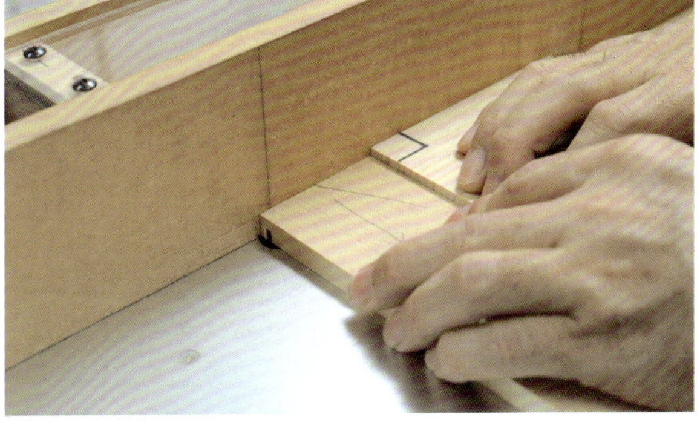

26 Create grooves in four corners of the side panels. The board width is narrow, so use backup material to stabilize.

PROCESSING THE FRONT PANEL GROOVES

27 Adjust bit the height for the front panel groove processing. Set to the same height as the side panel thickness.

28 Next, process the front panel. Place the inside of the front panel against the fence and proceed with processing.

29

Assemble the front and side panels. The right side is the front panel, and the left side is the side panel. Although the joint isn't yet complete, check how well they fit together. After this, I'll trim the front panel with a saw.

30 This is a DIY right-angle guide (see instructions at right). It's an excellent tool for straight cutting right angles with a saw.

31

Measure the distance from the bottom of the groove in the side panel to outside using callipers. Tighten the stop screw. This distance determines how much to trim off the front panel.

32

Clamp the right-angle guide to the workbench. Set the calliper measurement from step 31 on the front panel. The front panel will be trimmed in four places, so place a square rod as a stopper on the opposite side of the front panel. This will allow you to flip the panel and set it against the right-angle guide after cutting one side, keeping the calliper measurement the same.

◉ USEFUL DIY RIGHT-ANGLE GUIDE

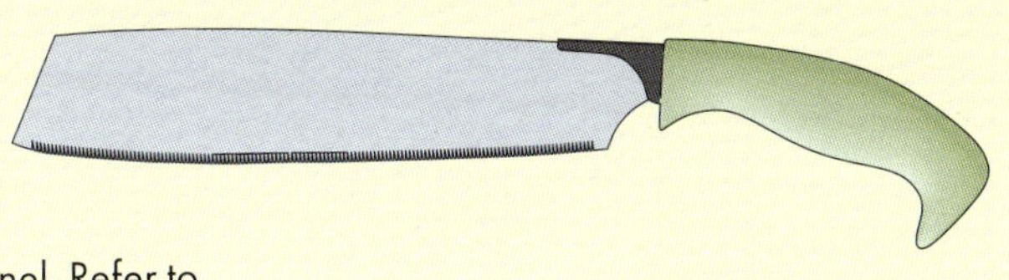

Use a saw guide with magnets to straight cut right angles. The saw blade sticks to the magnets, allowing for straight cuts. It's best to use a saw without set teeth. Here, a DIY right-angle guide is used to trim one side of the groove on the front panel. Refer to the diagram. This technique was devised by the author for making joints with minimal use of power tools. It is called the 'Sugita Method of Saw Woodworking'. It's explained in the books *Innovative Woodworking Techniques* and *Everything About the Sugita Method of Saw Woodworking*. Additionally, please refer to the commercially available aluminium right-angle guide on page 233.

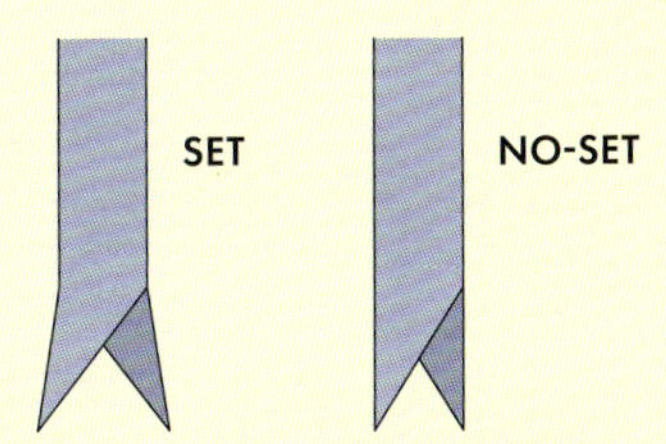

SUGITA METHOD OF SAW WOODWORKING

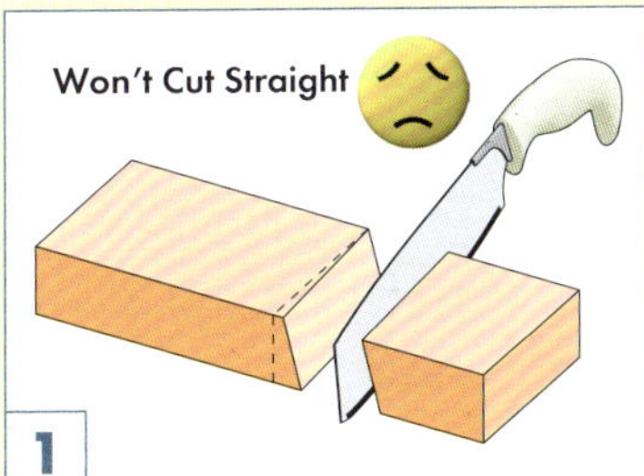

Cutting straight right angles with a saw can be quite challenging and requires practice.

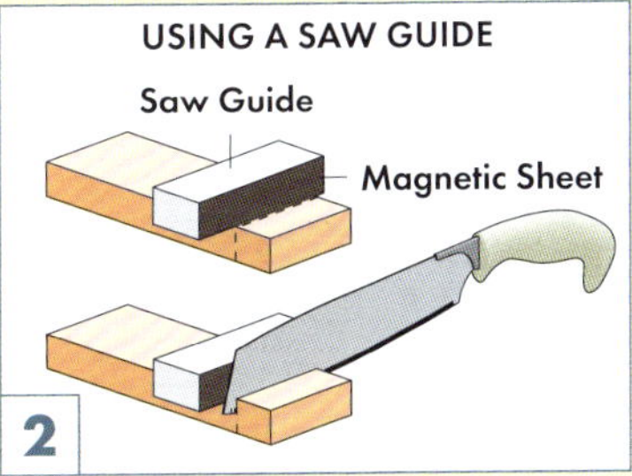

Use a magnetic sheet attached to a saw guide to hold the saw in place while cutting. Maintain a straight and right angle. It's cut at a right angle!

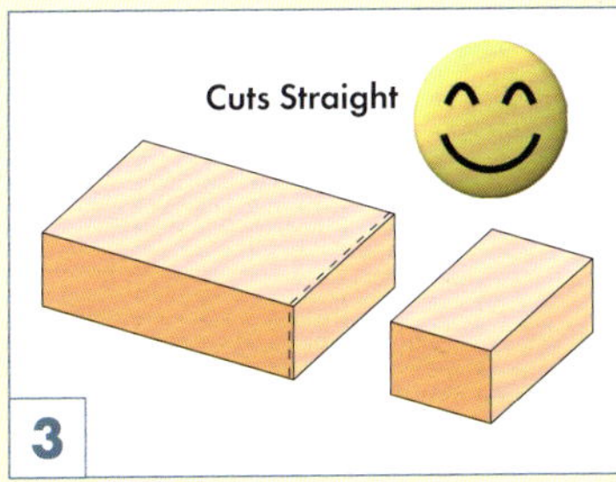

Using a saw guide makes it very easy to cut straight right angles. This is the 'Sugita Method of Saw Woodworking'.

DIY STRAIGHT RIGHT-ANGLE GUIDE

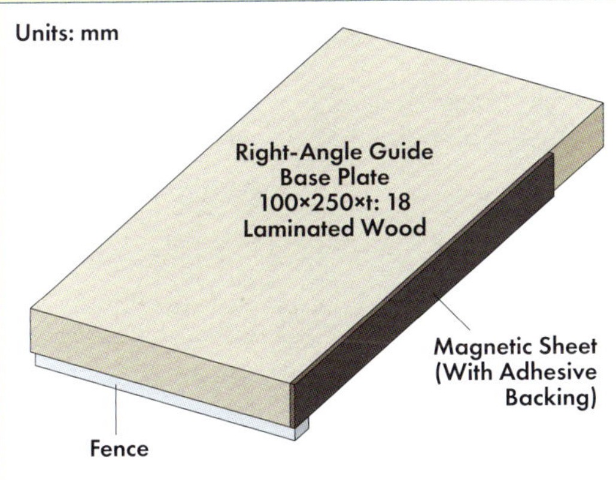

Cross-section: Thin square rod approximately 5×10. Glued perpendicular to the magnetic sheet. When material is placed against the fence and cut, a right angle will be produced.

Production steps

1. Attach a magnetic sheet to the base plate
2. Glue a longer fence to the magnetic sheet at a right angle (photo).
3. Finally, cut off the longer fence.

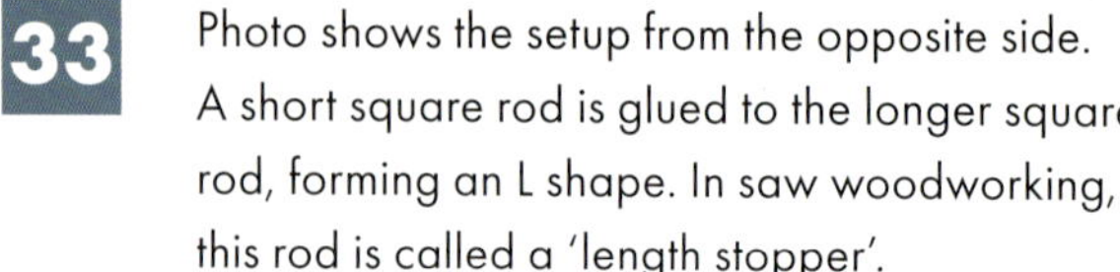

33 Photo shows the setup from the opposite side. A short square rod is glued to the longer square rod, forming an L shape. In saw woodworking, this rod is called a 'length stopper'.

34 Cut off part of the front board. This saw is a Craft 145.

35 The joints are now complete.

36 Dry fit the front and side boards and try sliding them into the opening. If they don't fit, adjust with a plane or sandpaper. Side boards should be about ¼in (5mm) shorter than the front board when dry fit, relative to the back board rabbet. Cut to this dimension.

PROCESSING THE FRONT AND SIDE PANELS (DRAWER JOINT)

First, cut dado grooves on the side boards and then determine the rear plate length. The width of the rear plate should also be fixed at this point. The method for processing the dado groove is based on the same principles as described on pages 42–43 and 138. Always perform test cuts on scrap wood before actual processing. The bit used is a ¼in (6.35mm) down-cut spiral bit. The cutting process exerts a slight lifting force on the material because of the spiral on the blade. If the material lifts, reprocess the same area to ensure a consistent depth.

37 The rear plate will be dadoed at ⅜in (10mm) from the rear edge of the side board. Use a spacer to set ⅜in (10mm) between the fence and trimmer bit. The groove depth should be set to ¼in (5mm). The back side of the fence should be processed as follows.

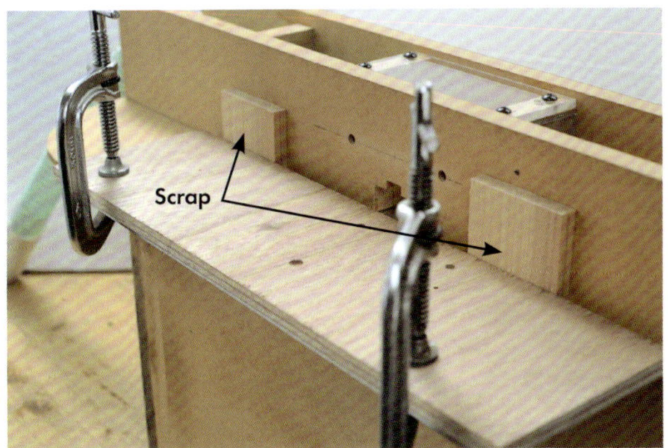

38 This photo shows the fence rear. The scraps on the rear plate are fixed on the plywood to allow us to perform the first dado cut.

39 Make the first dado cut. The side boards are narrow so they may become unstable during processing. Use backing material to stabilize them. Process all the side boards.

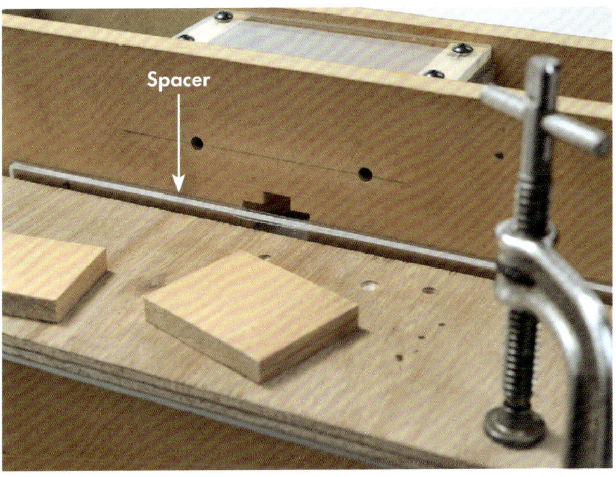

40

Next is the second cut to widen the dado. Release the fence and remove the scraps behind the rear plate. Don't remove the clamped plywood. This time, insert a spacer of the same thickness as the spiral bit diameter between the fence and plywood. Then, fix the fence. Now we are ready for the second cut. Perform a test cut with a scrap from the first dado cut.

41 This is a test cut with some scrap. When processed with a spacer of the same thickness as the router bit diameter, the fit was tight, so we reduced spacer thickness by 0.3mm and performed a third cut. Now, it's the perfect dado width. Use this setting for actual cuts.

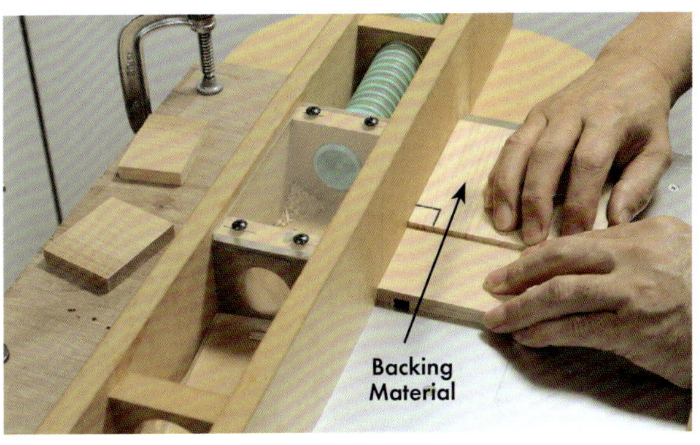

42 Perform the second cut. Again, we will use backing material during this step.

43

The dado grooves are complete. Any remaining splinters can be removed with sandpaper.

PROCESSING THE BASE PLATE GROOVES

Grooves will be cut into the front board and both side boards. The rear plate will not be grooved. The groove depth is set to ¼in (5mm). The groove width is about 0.5mm larger than the base plate thickness, allowing for a slightly loose fit. The bottom board will not be glued to the side boards. The method for widening the groove is the same as the dado process, using spacers and performing the cut in two passes. Since we are using 5⁄32in (4mm) birch plywood for the base plate, a ⅛in (3.2mm) down-cut spiral bit is used.

44 Grooving the base plate.

45 All groove processing is now complete. Small raised sections may become prone to chipping where grooves intersect, but this won't affect the structure, so there's no need to worry about it.

46 For the rear plate, the measure inner dimensions after dry fitting front and side boards. Cut the rear board to a length that includes dado groove depth on both sides.

47 Dry fit the rear plate. After this, cut the base plate and determine the rear plate width.

BASE PLATE

48 Dry fit, wrap in rubber bands, and measure the inner width. Subtract 3/64in (1mm) from the total measurement plus both base plate groove depths. Then, cut the bottom board to this size. The front-to-back measurement will be adjusted in the next step.

49 The base plate length in the front-to-back direction should be cut slightly long. Adjust the length as you check how well the drawer fits. Before taking measurements, flip the drawer over so the bottom faces up. Ensure the rear plate doesn't block the bottom board groove by adjusting its position. Then, insert a ruler into the front board's groove and measure the side board's rear edge distance. Add 3/8in (10mm) to this measurement and cut the bottom board to that length.

50 Insert the base plate from the rear of the drawer. Adjust the rear plate so it doesn't block the groove. Once the drawer is fully assembled, the rear plate will rest on top of bottom board. The rear plate height (or width) should be trimmed later to 3/64–5/64in (1–2mm) below the top edge of side boards.

51 Secure the base plate and rear plate with wood screws. Position screws at the centre of the rear plate's length and thickness.

GLUING

52

Apply woodworking glue and tighten the assembly using rubber bands or clamps. If any glue seeps into the base plate groove, clean with a cotton swab or similar. Also, use a square to confirm the inner corners are at right angles.

FINISHING

A Western-style plane is used for finishing. Western planes are easy to sharpen and use, thus providing excellent cutting performance.

53 Use the plane to smooth down the offset created on page 181, 'Drawer Structure and Construction Tips'.

54 Finish the front drawer opening to make it flat and smooth. The plane used is a 60½ block plane.

55 Plane down the overhanging part of the top plate to align with the side boards. The plane used is a No.4 Smooth Plane, Iron.

56 Use the front board of the drawer as a reference and shave down the slightly protruding side boards to ensure the drawer fits perfectly into the opening.

ATTACHING THE BACK BOARD

57 Cut out the back board and attach using wood screws. 5.5mm MDF is used here.

ADJUSTING THE DRAWERS

58 Adjust the front board position while planing the portion of the base plate that protrudes from the side boards. Here, the top drawer front board is sticking out slightly. If you plane the rear of the bottom board, the drawer will retract, aligning with the outer frame.

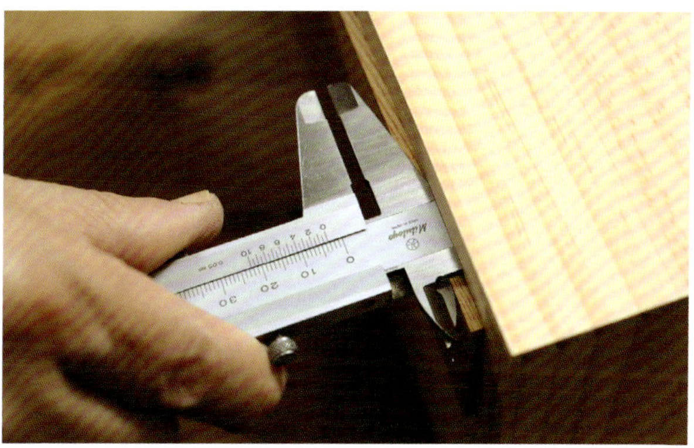

59 Use the sliding part of the callipers to measure the front board protrusion. Transfer that measurement to the base plate to determine how much material to plane off.

60 Roughly trim off the excess. Then, adjust using a plane or sandpaper.

61 Plane or sand the base plate until the front board is flush with the outer frame.

◉ DRAWER JOINT PROCESSING STEPS

This is a convenient method where both the front board and the side boards can be machined with a single fence setup. It's ideal for the joints between the drawer's front board and side boards.

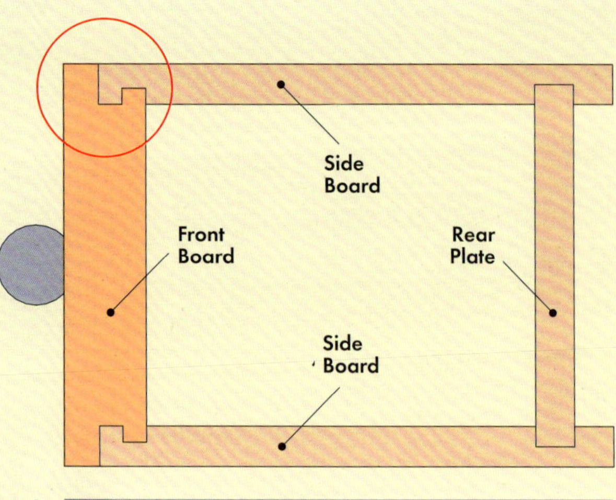

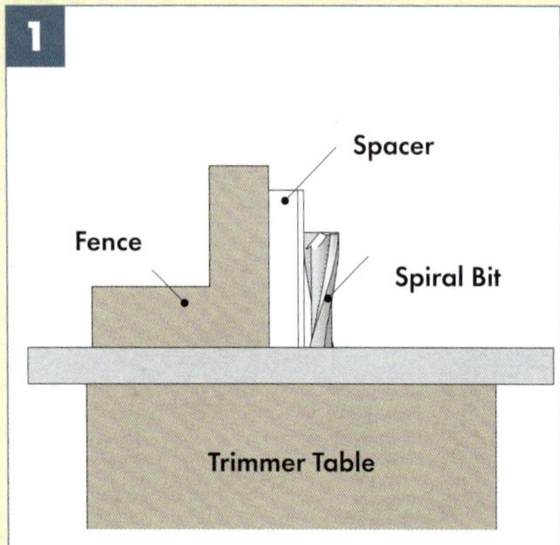

Create space between the spiral bit and fence. The distance should match the bit diameter. When setting the fence, use spacers like those used with the T-square slide fence (see page 40). Fully extend the bit.

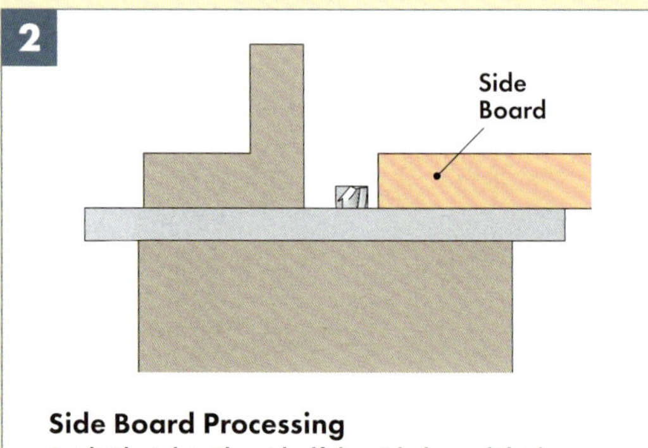

Side Board Processing

Set bit height. About half the side board thickness (as measured anywhere along the thickness).

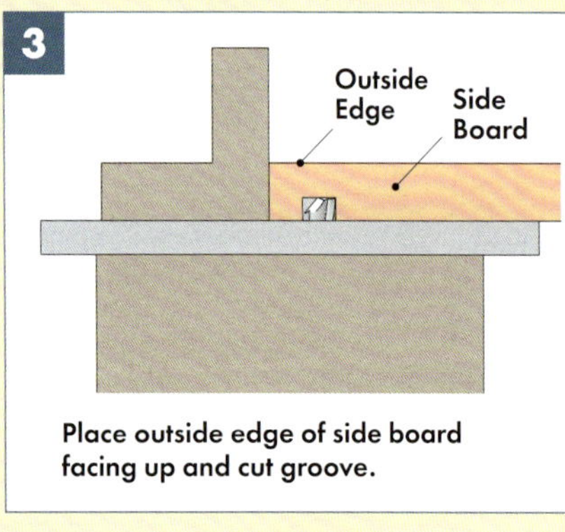

Place outside edge of side board facing up and cut groove.

4

Grooving is complete.

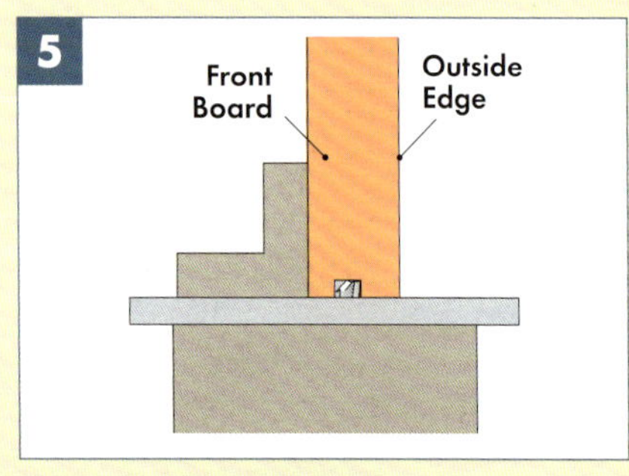

Front Board Processing

Cut first front board dado. At this point, position inner side against fence as shown in the diagram.

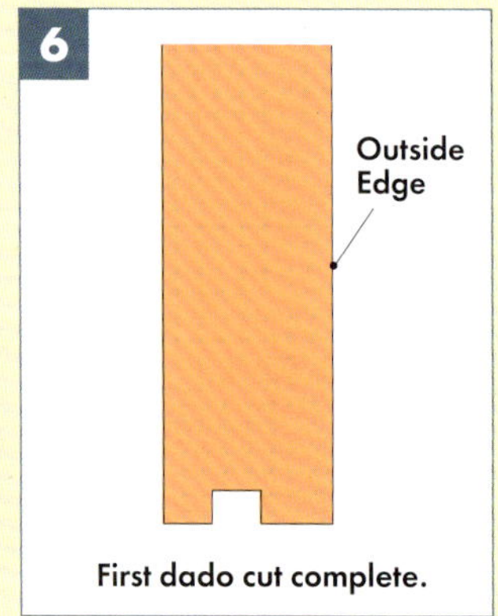

First dado cut complete.

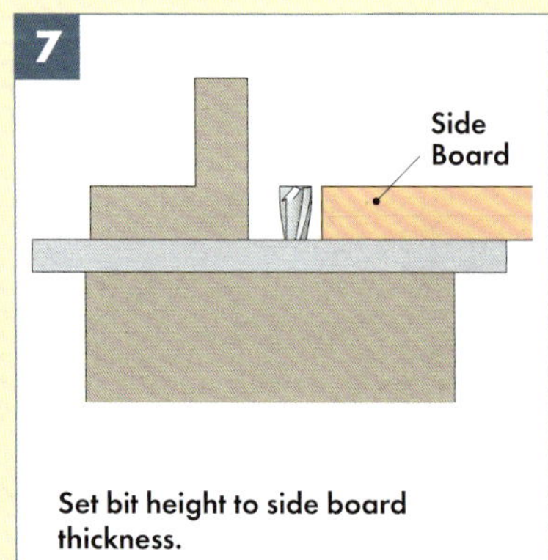

Set bit height to side board thickness.

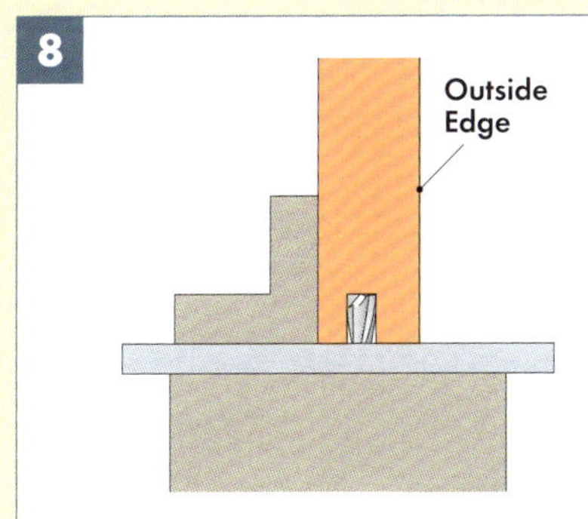

Perform second dado operation (front board dado is done in two parts, steps 7 and 8, to avoid putting too much strain on the bit).

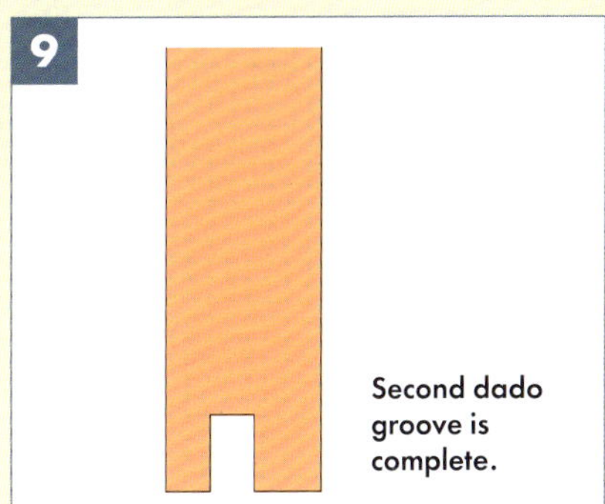

Second dado groove is complete.

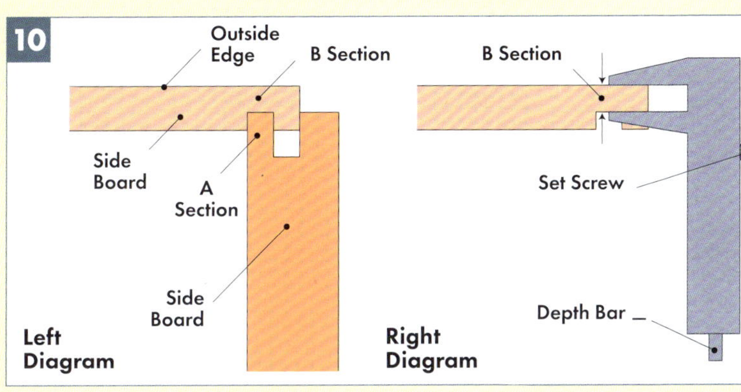

Dry fit as shown in the left diagram. Trim part A to complete joint. Trim amount to the same as part B thickness, where the side panel doesn't fit completely. Use callipers to hold part B, tighten the retaining screw, and fix it (there is no need to read the scale). Use callipers in that fixed setting.

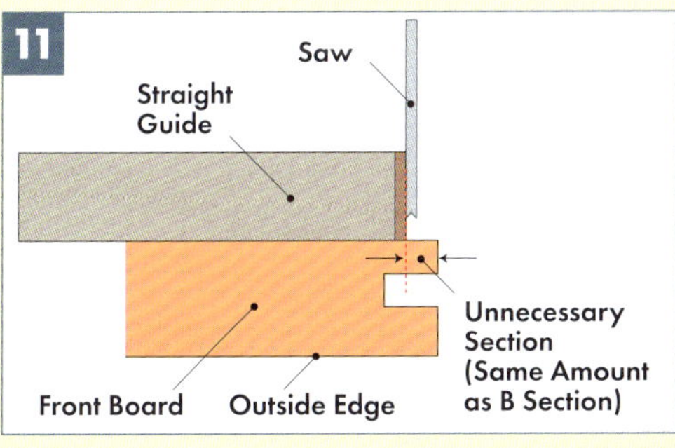

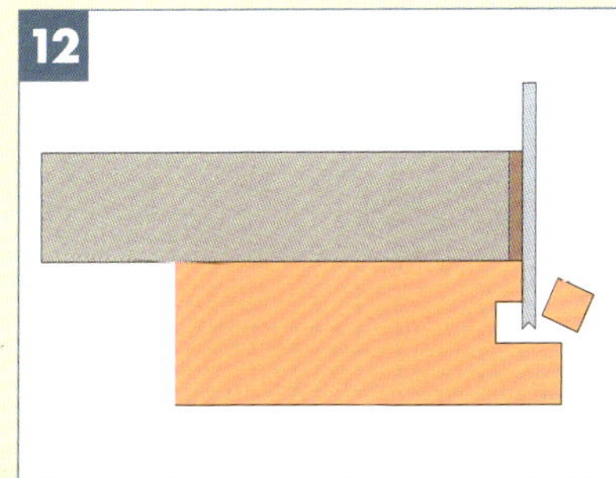

Use right-angle guide (see page 189). Cut off unnecessary portion of front board along the red dotted line.

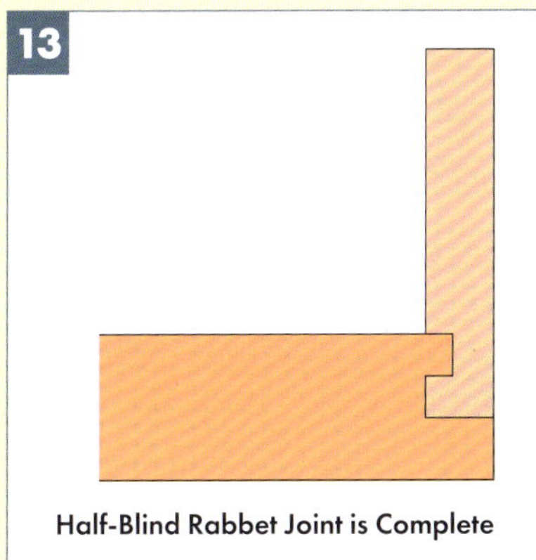

Half-Blind Rabbet Joint is Complete

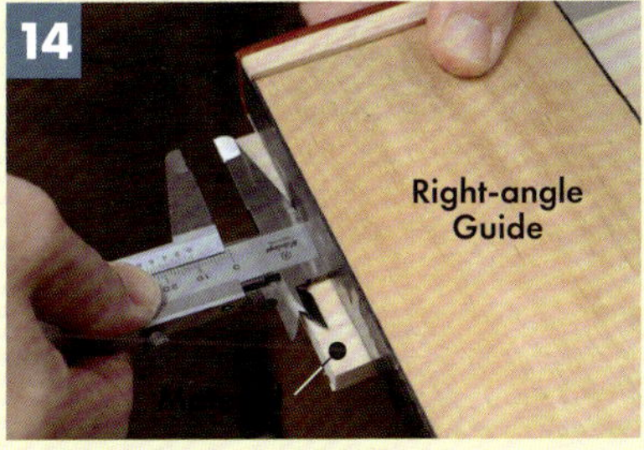

Setting Measurement With Callipers

Place the calliper top edge, fixed at the distance set in step 10 against the right-angle guide. Ensure the stepped portion of the sliding small jaw is on the back.

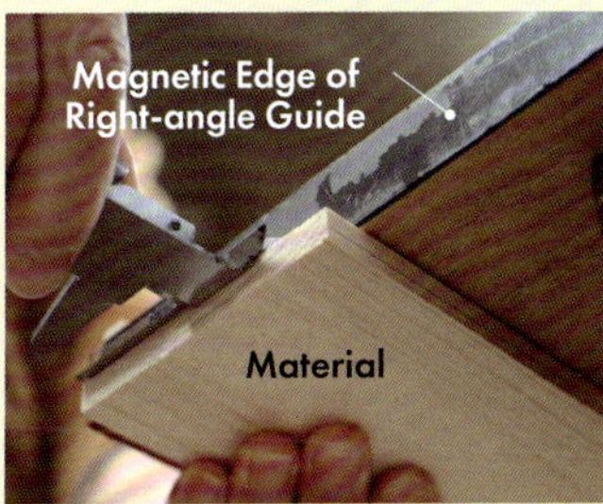

This is a view of the right-angle guide from below. You can see that the stepped portion of the sliding small jaw is pressed against the edge of the material. After this, clamp the material and cut it with a saw.

Photo Frame

Let's make a photo frame for a 5×7 picture. Creating steps that slope inward makes it easier to focus on the artwork.

The 45-degree mitre joint is known for often having gaps at the connection points and requiring high precision. Various electric tools, like circular saw tables or sliding saws, can make mitre cuts, but even if the angle display on the tool shows 45-degrees, they tend to be unreliable. It's common to make repeated test cuts and numerous adjustments. The author has gone through trial and error to find ways to achieve accurate cuts. The 'aluminium right-angle guide', which offers very high processing accuracy, was developed to address these concerns (see page 233). This jig is designed for cutting right angles, but it can also be used for 45-degree cuts. In this case, we will use the magnetic sheet on the right-angle guide. Please use a saw without a set (no angled teeth).

When making a mitre joint, the material must be cut to exactly 45-degrees, and both pieces must be the same length. If these two conditions are not met, joint gaps will appear. Let's start by ensuring that the material widths align.

PHOTO FRAME DIMENSIONAL DRAWING

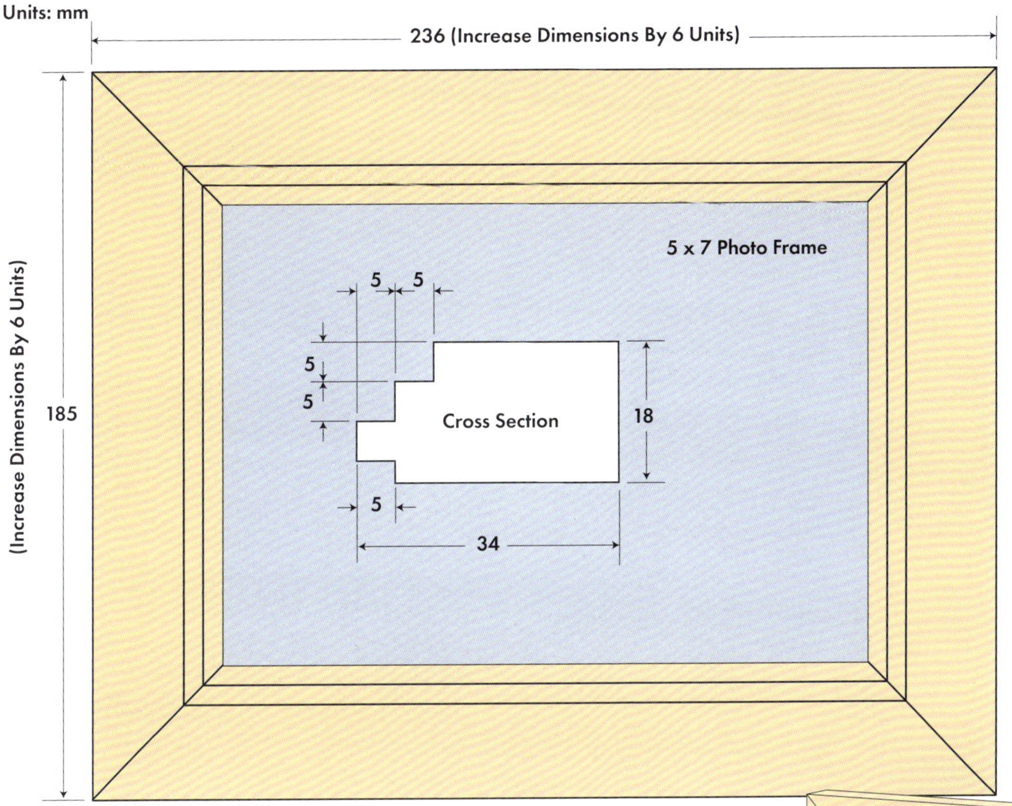

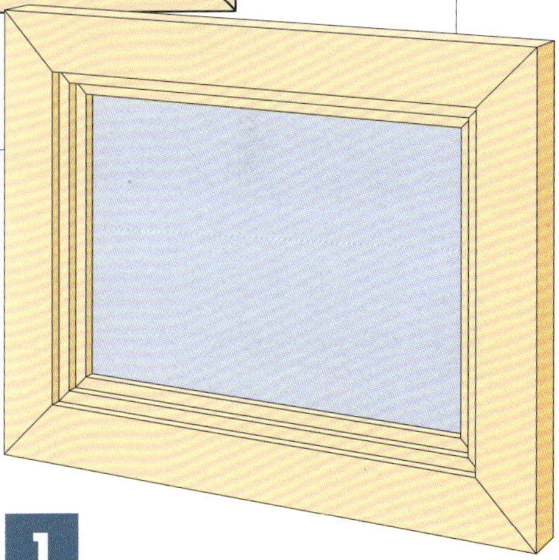

Units: mm

236 (Increase Dimensions By 6 Units)

185 (Increase Dimensions By 6 Units)

5 x 7 Photo Frame

Cross Section

5 5

5

5

18

5

34

CUT MATERIAL TO SAME SIZE USING STRAIGHT GUIDE

1

Use the aluminium right-angle guide (see page 233) to trim a small portion off the material end grain. This ensures the edge is square and removes any areas that may have collected small stones while the material was stored in an upright position.

2

Trim the material to ¼in (6mm) longer than the finished dimension. Once the material is clamped to the right-angle guide, place an L-shaped block made from scrap, referred to as a 'length stopper', at the material's end. This helps cutting each piece to the same length. Once the guide is clamped in position, loosen the material clamp, reposition it against the length stopper, and clamp again. The length stopper should be made from a scrap piece that is at least 9¾in (250mm) long.

3 The magnetic force of the sheet is strong enough to guarantee that the material will be cut straight and at a right angle.

4 Two pieces of material have been cut. Cut the remaining two pieces in a similar manner.

45-DEGREE CUTTING

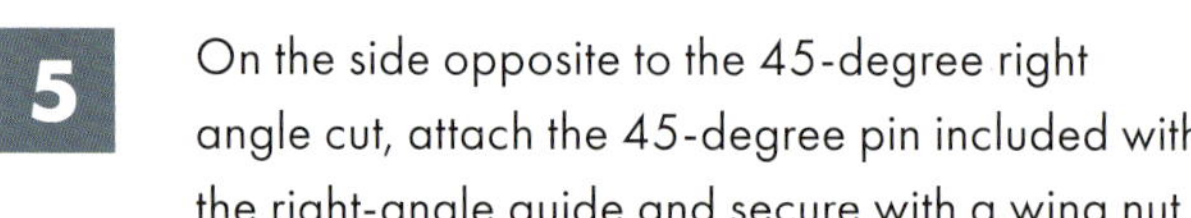

5 On the side opposite to the 45-degree right angle cut, attach the 45-degree pin included with the right-angle guide and secure with a wing nut.

6 Clamp the right-angle guide to the workbench. The aluminium fence extends downwards on the right angle cutting side. Clamping directly may cause deformation, so add scrap material that is at least 5⁄32in (4mm)-thick underneath before clamping. Also, use the workbench corner to secure the guide at approximately 45-degrees.

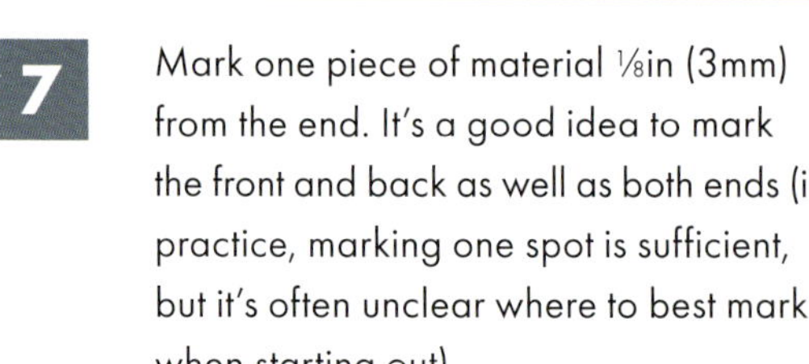

7 Mark one piece of material ⅛in (3mm) from the end. It's a good idea to mark the front and back as well as both ends (in practice, marking one spot is sufficient, but it's often unclear where to best mark when starting out).

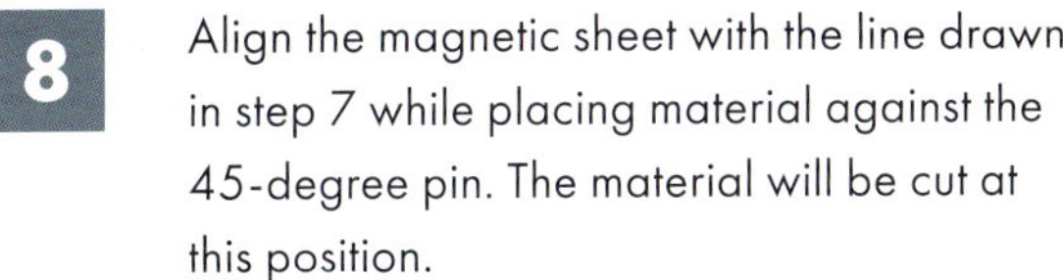

8 Align the magnetic sheet with the line drawn in step 7 while placing material against the 45-degree pin. The material will be cut at this position.

9 Clamp the length stopper securely to the workbench while positioned against material end. Be careful, if the length stopper shifts even slightly, accurate processing will be impossible.

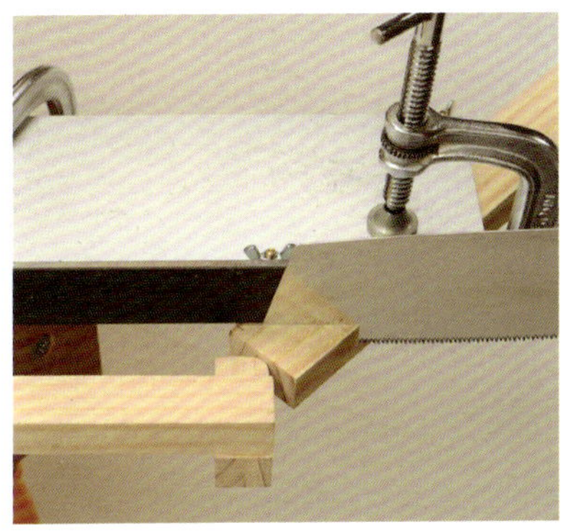

10 Cut the material. Use short strokes as there is limited space to move the saw blade.

11 After cutting, remove burrs from the end grain with a sanding block or similar.

12 Flip the material over and re-clamp it against the length stopper.

13 Cut 45-degrees. When flipping the material, ensure that the already cut end is facing the direction circled in the photo.

14 Dry fit using rubber bands. There should be no gaps.

FINISHING THE INSIDE EDGES

15 It's easier to finish interior surfaces now rather than after assembly. Use sandpaper or a planer to smooth surfaces.

DADO CUTTING STEPS

16 The router bit being used is a straight bit with a diameter of ¾in (19.05mm).

SETTING TRIMMER BIT POSITION

Two methods will be introduced here. One method is to align the bit's cutting edge with a pencil line, and the other uses thickness spacers.

17

The dado will have a stair-step shape, measuring ¼in (5mm) in both dimensions. Mark a pencil line on scrap. Align the cutting edge of trimmer bit with the line. Set the distance from the fence and trimmer bit height.

18 To determine the fence to bit distance, place a ⅜in (10mm)-thick acrylic spacer against the fence, and an acrylic sheet is used as a ruler. Move the fence until the ruler touches the bit tip. This sets the distance to ⅜in (10mm).

19 To set the trimmer bit height, combine various acrylic spacers to reach the desired thickness. Then, place the thicker acrylic sheet on top to use as a ruler. Raise the bit until it touches the ruler to set to the desired height.

20 Once the bit position is set, proceed with dado cutting. Start by making cuts a few millimetres high.

21 Gradually increase the height as you continue cutting until you reach the final desired height.

22 Make the second step. For safety, keep your fingers as far away from the fence as possible while holding the material.

23 Stair-step dado cutting is now complete.

25 After completing cuts on both sides, glue.

PROCESSING THE BACK SIDE

24 This dado is for inserting the transparent back panel and the photo backer board. The dado depth should equal thickness of transparent panel plus thickness of backer board.

GLUING

26 Use rubber bands to temporarily assemble the frame and check the overall fit.

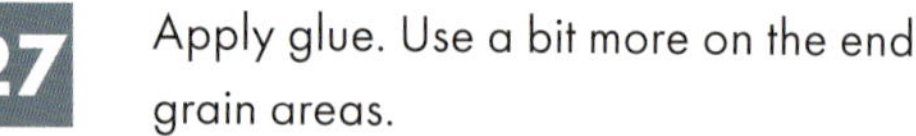

27 Apply glue. Use a bit more on the end grain areas.

28 Use thick rubber bands to increase the clamping pressure.

TRANSPARENT HOLDER AND BACKER BOARD

29 Measure the internal dimensions of the dado on the back side. Cut the transparent panel and backer board 3/64–5/64in (1–2mm) smaller than that.

30 The backer board is cut from 1/8in (3mm) plywood. One side should be white.

31 The transparent panel is cut from 3/64in (1mm) PVC sheeting. The yellow tool is an acrylic cutter.

FINISHING

32 Sand the frame or use a planer for finishing. The material under the frame is a non-slip mat.

33 Finish thick areas in the same manner as in step 32.

34 Finally, chamfer all corners with a planer or sandpaper.

COATING

35 Apply coloured varnish (gel coloured varnish) for the final finish.

INSTALLING HARDWARE

36 Secure the holding brackets with screws on four sides.

37 You can attach a hanging string by inserting pushpins at three locations: left and right sides and top. The top pushpin will be used to hang the photo frame on the wall, but for now, it's just temporary. If you firmly insert pushpins, while wrapping the string around the needle, the string will be compressed and won't come undone. And that's it! The string will be invisible when hung on the wall.

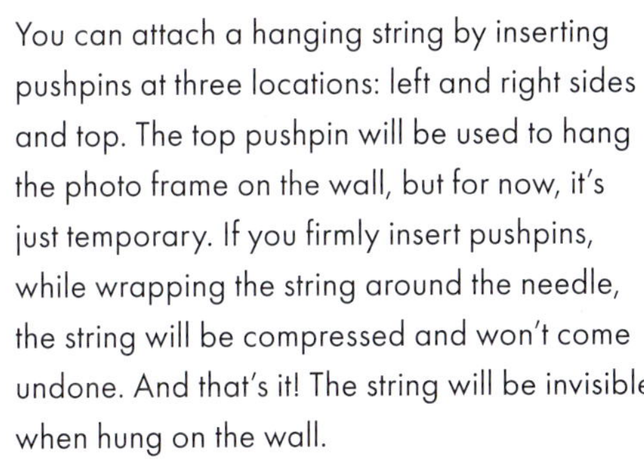

Wall-Mounted Box

Here, let's try out mitre and spline joints for grooving. For the mitre joint, a jig inclined at 45-degrees is essential. See page 176, 'Method for Producing 45-degree Inclined Board', for instructions. French cleats are used for hanging.

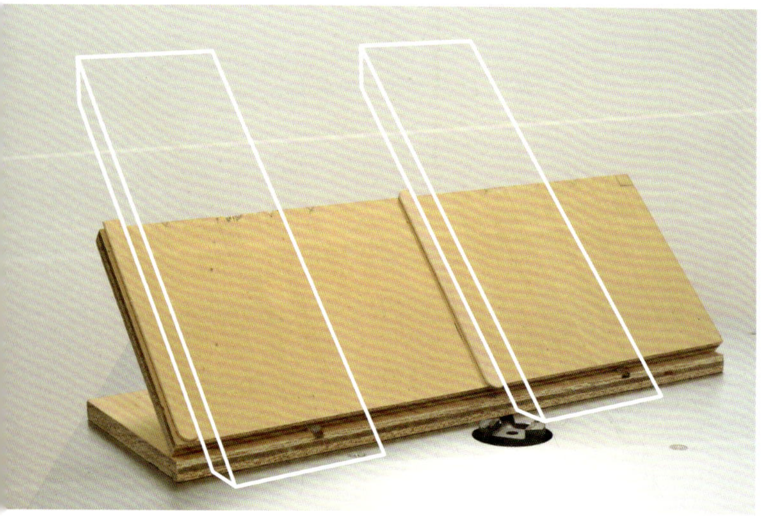

◉ PRODUCTION STEPS

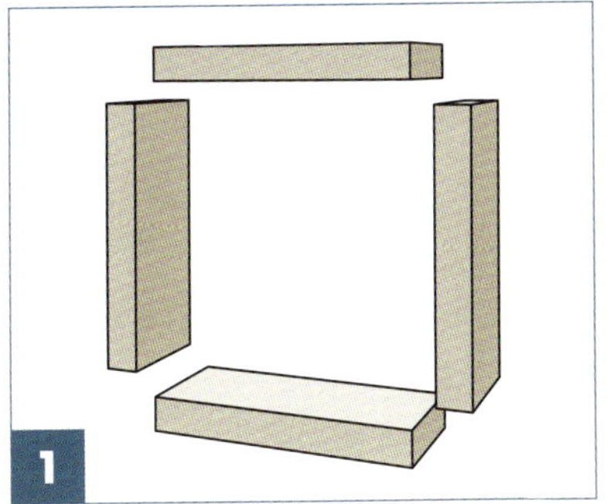

Cut material to size.

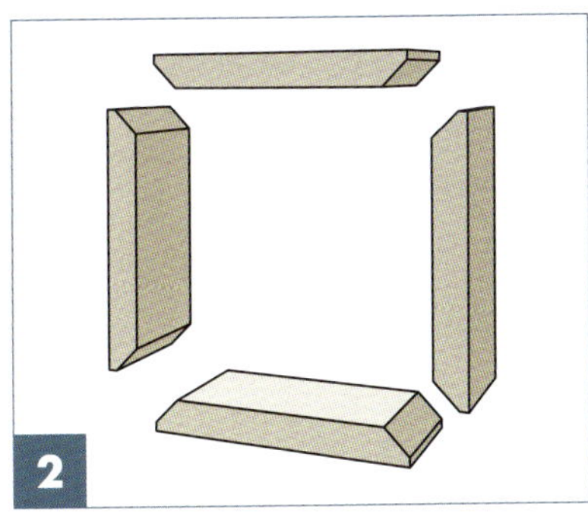

Roughly trim both ends at 45-degrees.

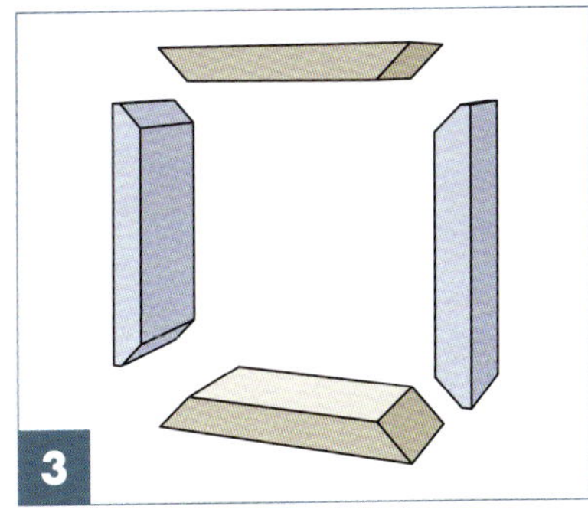

Use the trimmer table to cut mitre joints in the first two pieces together.

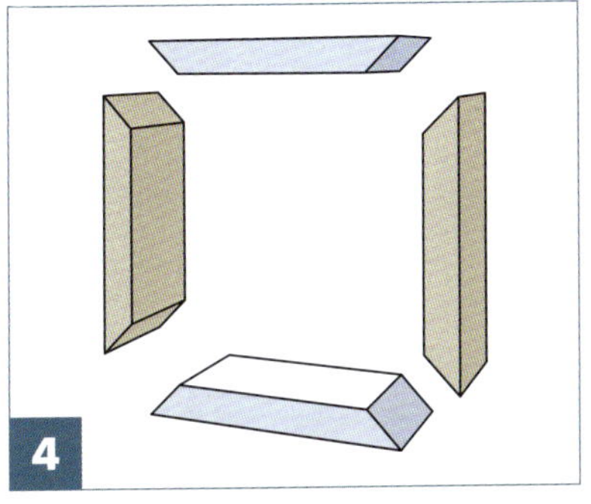

Cut mitre joints on the remaining two pieces together.

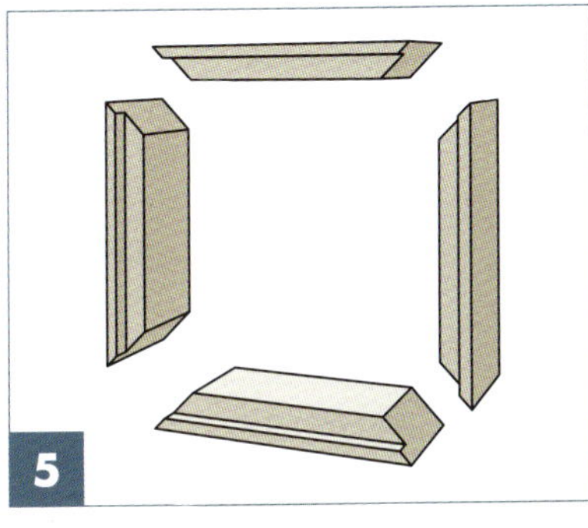

Machine the dadoes for inserting the rear plate.

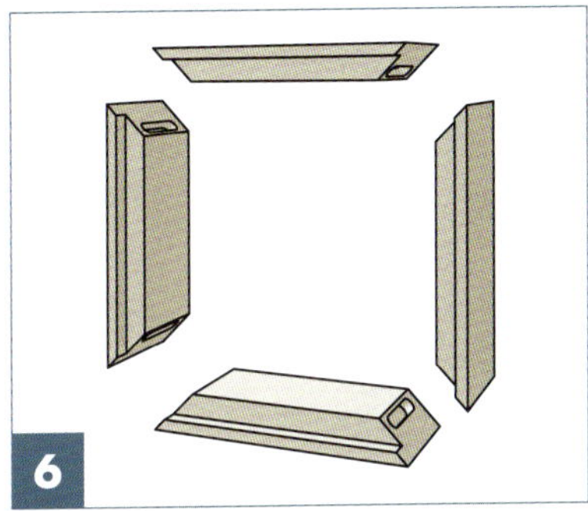

Cut the spline joint grooves.

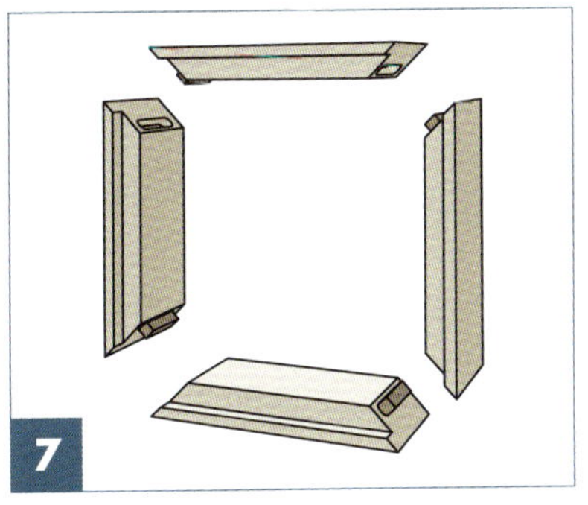

Apply glue and insert splines in grooves.

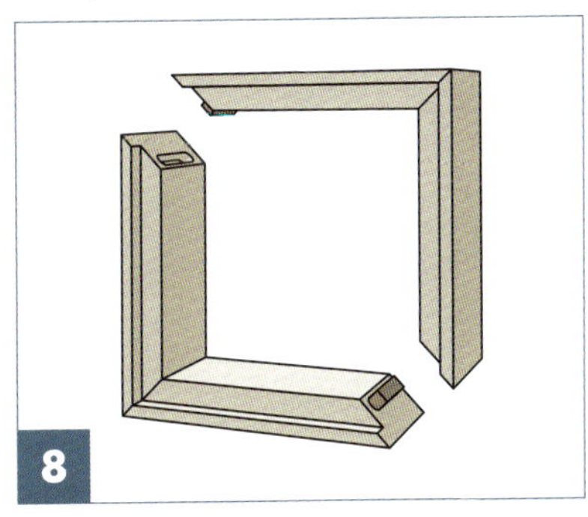

Assemble everything and tighten with belt clamps.

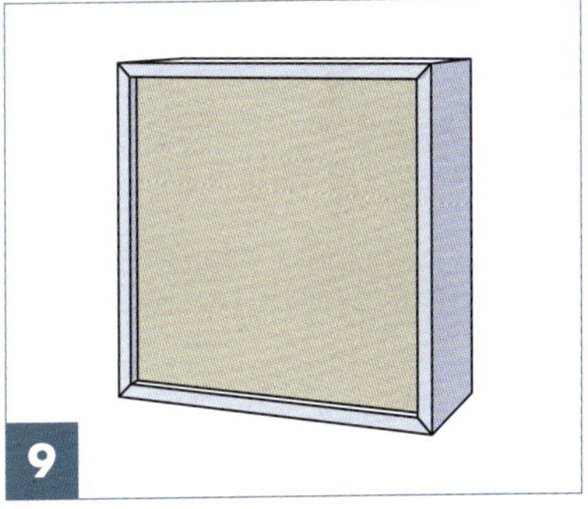

Cut out the rear plate, insert, and secure with screws.

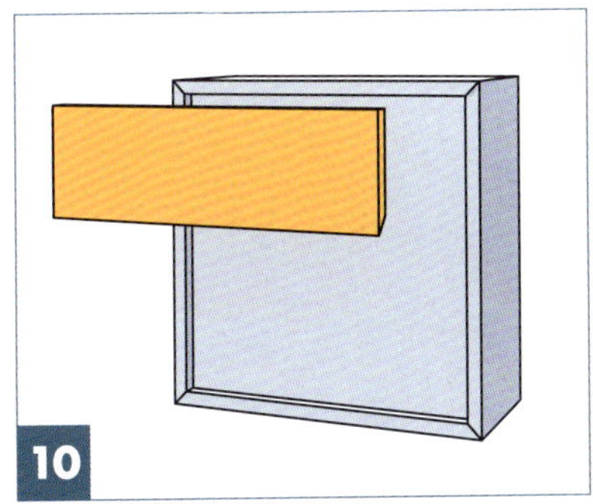

Cut the French cleat and secure with screws.

CUTTING MATERIALS

First, cut material to the required length and create mitre joints on both ends. This processing is done on a trimmer table. It's better to roughly cut the ends at 45-degrees with a saw beforehand to reduce the load on the router bit. For this, a simple saw jig is used. The jig shown in the picture is used for this purpose. Behind it are materials cut using this jig.

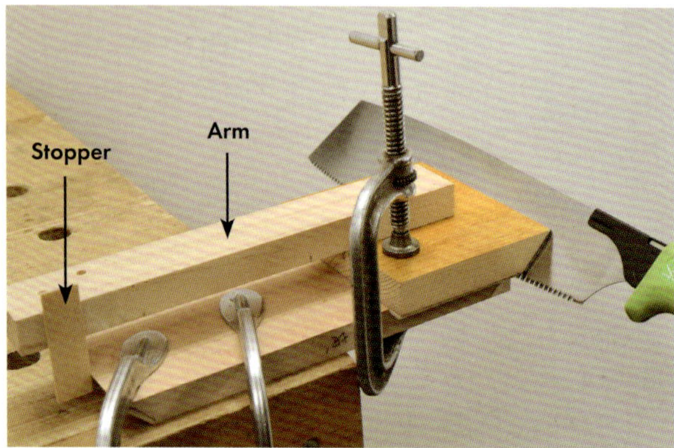

2 A 3/4in (20mm)-thick piece of scrap is cut at 45-degrees, and a saw guide with a magnet sheet is used. The arm and stopper are fixed to the jig using double-sided tape. The jig is roughly 45-degrees. It's okay if the angle isn't perfectly precise.

3 The material is clamped to the workbench, and the jig is clamped on top. The dotted area shows where the material has already been cut off. The key here is where to cut the material. Instead of cutting entire material thickness at a diagonal – like a planer blade – leave a 3/64–5/64in (1–2mm) flat section at tip and cut. This flat part allows the stopper to rest on it when cutting the opposite side, making it easier to cut to the same length.

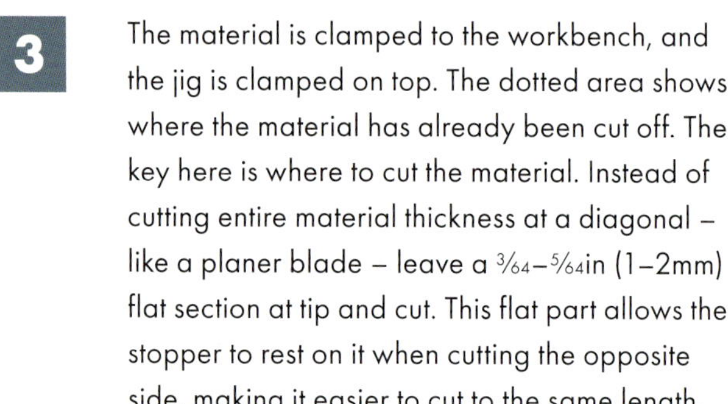

4 Now, trimmer processing can begin. Mark the top, bottom, left, and right sides. By processing the top and bottom, and left and right together, you can make identical-sized parts. To do this, tape the two pieces together with packing tape. When taping the materials together, put outer (visible) edge grain sides together. This prevents chipping during processing.

PROCESSING MITRE JOINTS

5 Tape the top and bottom, left and right pieces together, creating two pairs. Clamp these pairs in the mitre jig. Both sets should be clamped with a gap of 3⁄64–5⁄64in (1–2mm) between the materials and workbench surface.

6 Rear view of the jig. While slightly different from the structure explained in 'Method for Producing a 45-degree Inclined Board' on page 176, it is based on the same concept.

7 For this step, use a large-diameter straight bit with a short cutting edge to remove large amounts of material at once.

8 Slightly raise the bit and begin cutting. After each pass, flip the material and check for any areas that remain uncut. If you remove too much material at once, it can cause chipping.

9 Gradually raise the bit until all four surfaces are evenly trimmed with no missed areas. After this, the first pass is complete. Remove clamps (but don't remove the packing tape yet).

10 For the other side, hook a measuring tape onto the completed mitre-cut end and mark the desired length with a pencil.

Align the bit height with the pencil mark you made in step 10. Adjust the trimmer hose clamp to this setting as well. That way you can gradually reach the desired depth after just a few passes.

Process top/bottom and left/right materials, one set at a time.

After processing both sets of materials, the surfaces are neatly finished. Now, peel off the packaging tape.

14 Dry fit using rubber bands and check joints. All joints should fit tightly, with no visible gaps, as shown in the photo.

PROCESSING DADO FOR BACK PANEL

15

The next step is to cut a dado for inserting the back panel. A French cleat will be used for wall mounting, so you'll need to account for the combined thickness of back panel and cleat. The back panel is 3/32in (2.7mm) plastic, and the French cleat is 1/4in (5.5mm) plywood. The combined measurement is about 5/16in (8mm). This measurement will be used for the dado.

The bit being used is a 3/4in (19.05mm) straight bit.

17 Set the fence distance and cutting depth using plastic spacers. This is a dado cut, so the fence is set up on the dust collection port side. Set the maximum cutting depth using the trimmer's hose clamp.

18 Make multiple shallow passes, removing a few millimetres of material at a time.

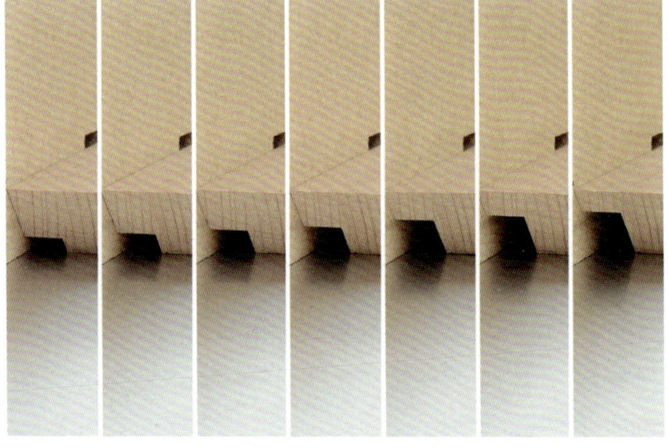

19 Continue with several passes, as shown in the photo, until you reach the desired depth.

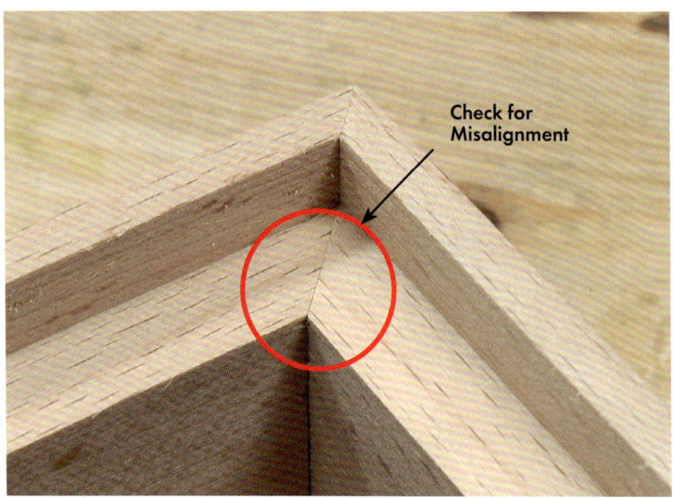

20

After dry fitting, check there are no misalignments in the dado sections. The next step involves creating grooves to reinforce the 45-degree angles with a spline joint.

◉ REINFORCING WITH A SPLINE JOINT

45-degree joints involve end grain gluing, which is inherently weak, so reinforcement is necessary. To achieve this, grooves will be cut for splines, which will add strength. For details on jig used, see page 45 'Edge Guide Jig for Spline Joints'. The 45-degree jig and worktable (the outlined portion in the photo) are used to process mitreed material. The edge guide jig is used with a trimmer to determine groove width.

SPLINE JOINT GROOVE CUTTING

21

The edge guide jig attached to the transparent housing. This setup allows for spline groove cutting on the 45-degree mitreed edge. The bit is a ⅛in (3.2mm) up-cut spiral bit. The jig features two slits (pockets) visible on either side. Inserting spline offcuts as spacers into these slits allows you to adjust groove width and perfectly fit your splines. This is an idea patented by the author.

22 As preparation, cut two spacer pieces from the 5⁄32in (4mm) plywood that will be used for the spline.

23 Clamp the 45-degree jig securely in a bench vice or similar.

24 Attach the worktable to the jig.

25 Tighten the jig firmly in place using wing nuts.

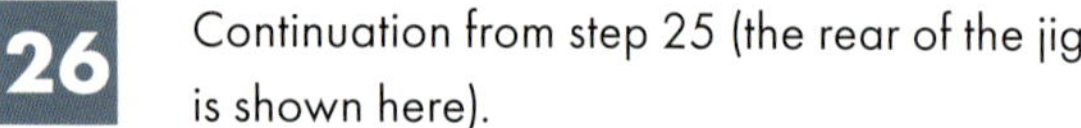

26 Continuation from step 25 (the rear of the jig is shown here).

27 Widen the edge guide jig and insert spacer pieces cut in step 22 into jig slits.

28 Position the workpiece mitred edge against the jig. Align carefully while clamping securely.

29 With the edge guide jig against the worktable, determine the groove's location and depth before cutting. This is the first groove pass. For better results, place the groove as close to inside the material's as shown in the previously processed piece (bottom left of this photo).

30 Unlike a router, we can't adjust the bit height during operation, so start by tilting the trimmer slightly so the bit doesn't contact the material. Once the position is set, switch trimmer on and gradually return it to a horizontal position, allowing the bit to cut into material smoothly.

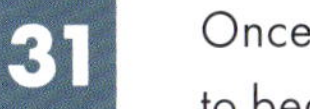

31 Once the trimmer is horizontal, push it forward to begin groove cutting.

32 When the groove reaches the desired position, turn off the switch while maintaining its current position to prevent any unevenness.

GROOVE WIDENING

33

Remove the spacer pieces that were clamped in the edge guide jig. This allows you to adjust the bit's position to widen the groove. Tilt the trimmer so the bit doesn't touch the material, then turn on the switch.

34 Once the trimmer is horizontal, move it forward.

35 After reaching the desired position, turn off the switch to finish widening.

GLUING

36 Before gluing, prepare the material's inner surfaces. It will be difficult to finish these areas after assembly, so it's best to do it now using sandpaper or a planer.

37 Apply glue and insert the wedge into the groove at a 45-degree angle. Then, assemble the entire structure and glue together. Use a belt clamp or similar tool to tighten securely, ensuring that there are no gaps at the joints. Also, check for square.

COATING

38

Dilute Japanese sumi ink with water and apply. This will enhance the wood grain while giving it a dark colour. Subsequently, apply sanding sealer and a matte clear lacquer as described on page 144 for wall-mounted shelves.

39 Attach the back panel.

40 The back panel will be made from white polyester laminated plywood. After this, make a French cleat for wall mounting (see pages 122–123).

FRENCH CLEAT

41 Use a French cleat to hang on a wall. Two cleats, each cut at a 45-degree angle will be used here.

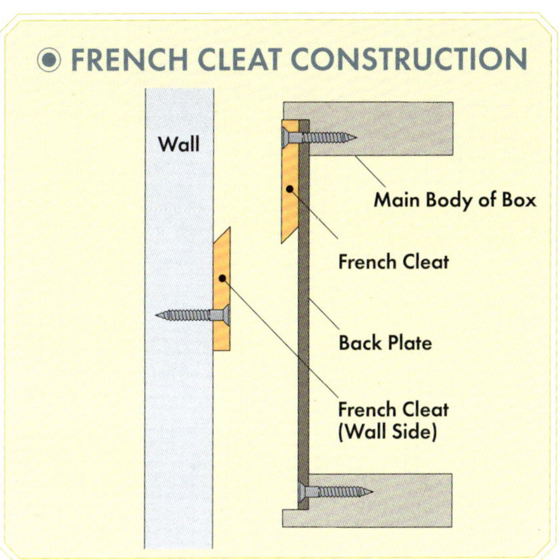

42 Use a 45-degree bit. It can be either a bearing-type bit or not, as shown in the photo (left: 90-degree V-groove bit; right: corner radius bit).

43 Set the bit height and fence position so that a small flat area remains at the tip of the material after the 45-degree cut.

44

When aligning the material against the fence for the 45-degree cut, having a slight flat area at the tip ensures stability. If the tip is sharpened like a plane blade, a gap will form between the processed part and fence, leading to instability as the unprocessed area decreases. This makes it difficult to achieve a perfect straight 45-degree cut.

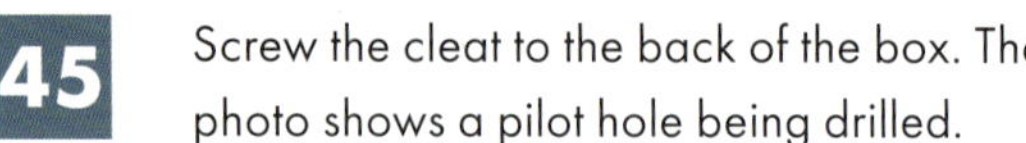

45 Screw the cleat to the back of the box. The photo shows a pilot hole being drilled.

46 The cleat is now attached.

◉ FLUSH PANEL DOOR

The flush construction method involves creating a frame from core material and sandwiching it by adhering plywood or similar materials on both sides. This method is used for furniture doors and interior doors. The interior is hollow, which reduces weight and makes it easy to construct. Start by marking a pencil line on the plywood at the desired panel size. Then, glue core material along that line. After that, use a flush trim bit with a bearing at the tip to trim any excess plywood.

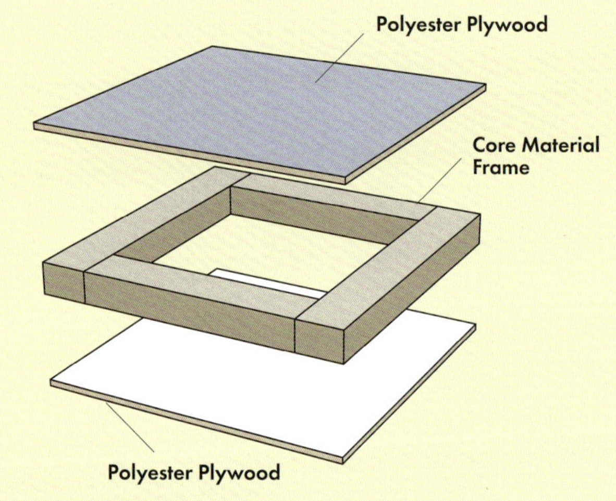

◉ FLUSH PANEL FABRICATION STEPS

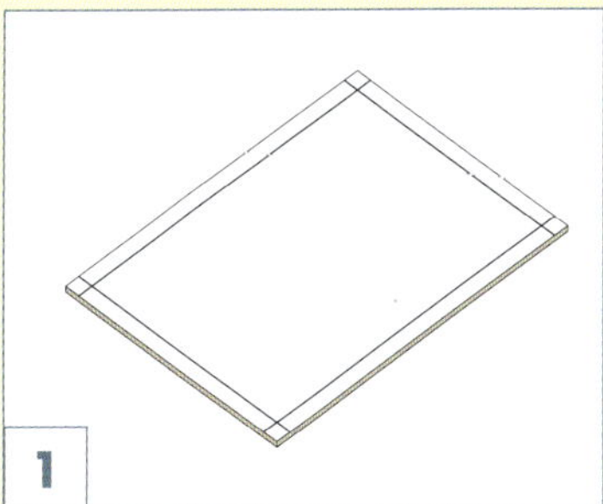

Mark the panel dimensions on plywood with a pencil.

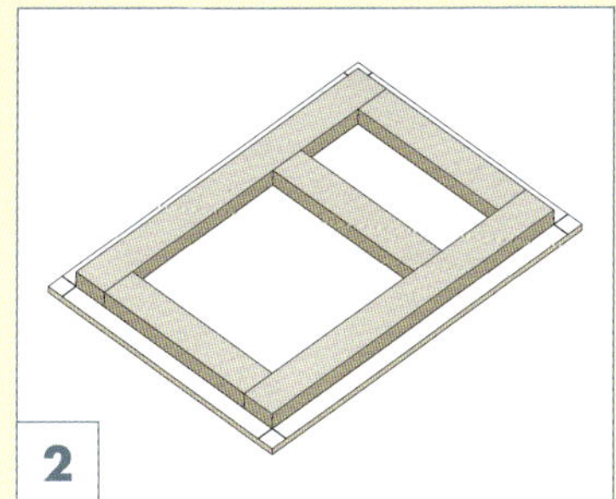

Align the core material along the pencil lines to a create the frame. Glue in place.

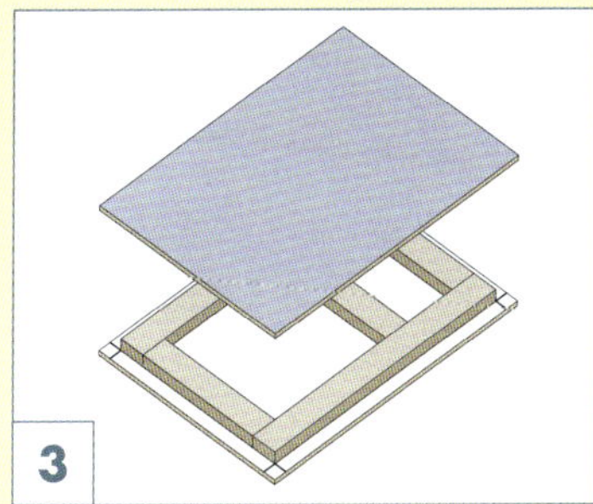

Adhere the plywood to the top of the core material.

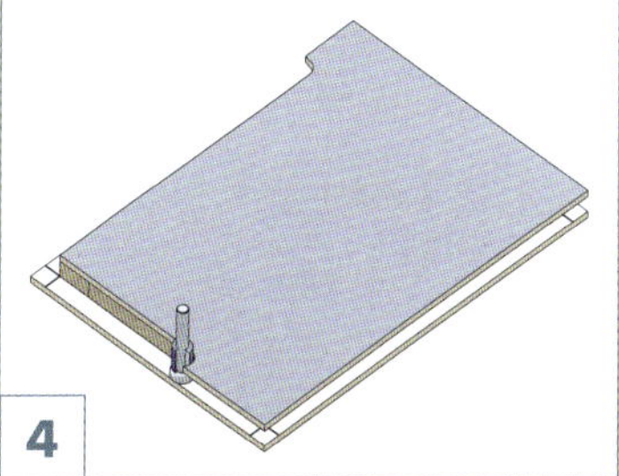

Set the flush trim bit depth so that the bearing part makes contact with the frame. Trim off any plywood that extends beyond the frame.

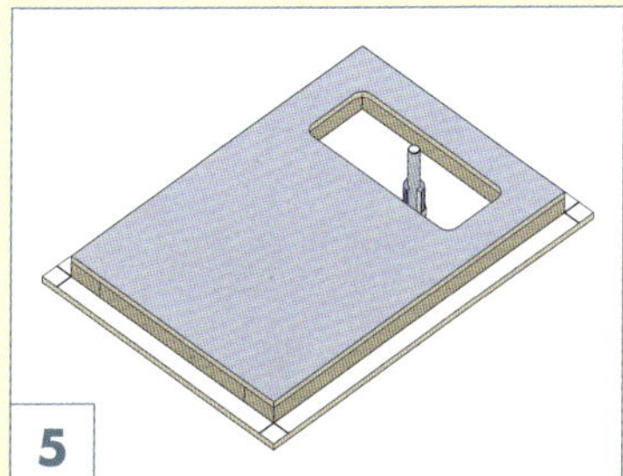

Process the window openings in a similar fashion (drill the holes beforehand and insert the bit to begin cutting).

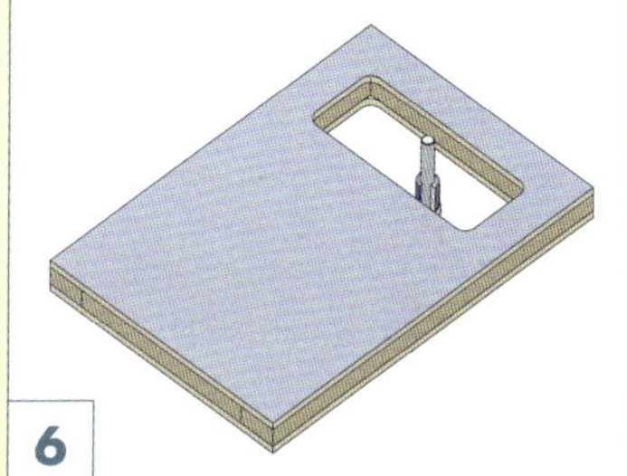

Flip the panel upside down and process the opposite side in the same manner (descriptions on pages 220–221 use a trimmer table, so the orientation is reversed).

MAKING FLUSH PANEL

47 Accurately measure the opening where the flush panel door will be installed. The panel should be made a few millimetres smaller than this measurement on all sides.

48 Mark the door dimensions on the plywood back. Glue the core material along the marked lines to create the frame. Even slight differences in the core material thickness can cause gaps or unevenness in the panel surface, so cut it from a single sheet of plywood.

49 Ensure the frame joints are as gap-free as possible. Gaps can lead to indentations on the panel edges during trimming.

50 Apply glue evenly over the entire surface. If the door is large, add cross-shaped or other reinforcements on the inside of the frame for additional support.

51 Place another sheet of plywood on top and glue it down.

52 Use multiple spring clamps or similar tools to hold everything in place while gluing. Ensure there are no gaps between panel and frame.

53 Make sure there are almost no gaps at the frame joints.

54 Adjust the flush trim bit height so that the bearing part makes contact with the frame.

55 As you work from left to right, trim any plywood that overhangs the frame. Rotate the panel as indicated by arrows while processing.

56 The flush trim bit bearing diameter is the same as the cutter diameter. The panel will be shaped to the same size as the frame.

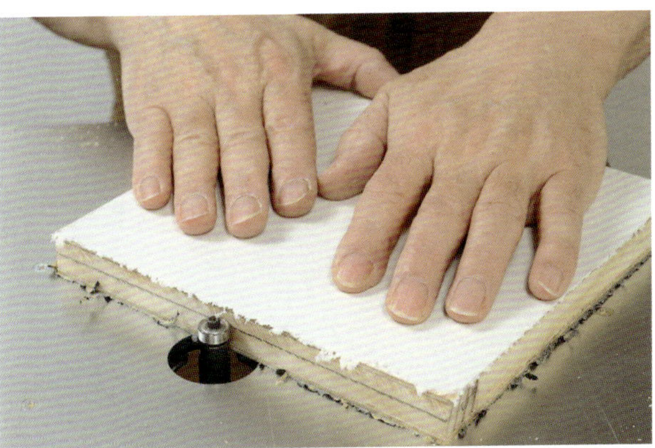

57 Shape the opposite side as well. Sometimes, pieces of polyester sheet adhered to the plywood surface may go untrimmed. After shaping, remove with sandpaper or a plane.

58 The flush panel is complete. Attach hinges, handles and stoppers.

Cutting Board

In this section, you will learn two techniques. First, you'll learn how to laminate three pieces of material together, eliminating any offset or unevenness and processing them to a uniform thickness. Second, you'll learn how to use a template and template guide for groove processing. This process typically utilizes the plunge function of a router. If you find that challenging, you can complete the project without grooves (see page 26 'Template Guide').

For details about the router's plunge function, see page 24 'Differences Between Trimmers and Routers'. Additionally, for information on processing material to a uniform thickness, see page 36 'Trimmer Planer'.

◉ CONSTRUCTION STEPS

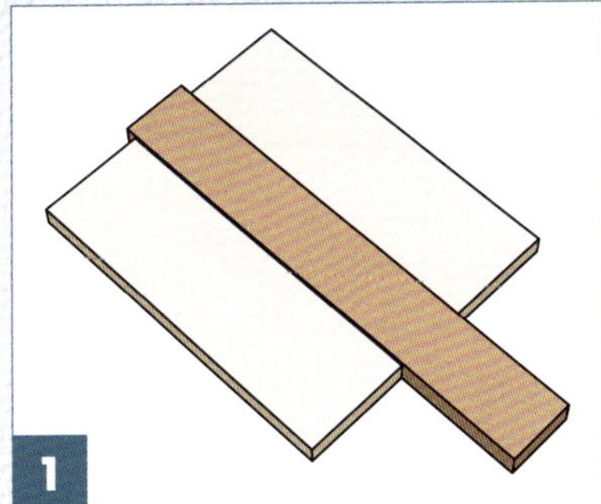

Laminate three pieces of the material together.

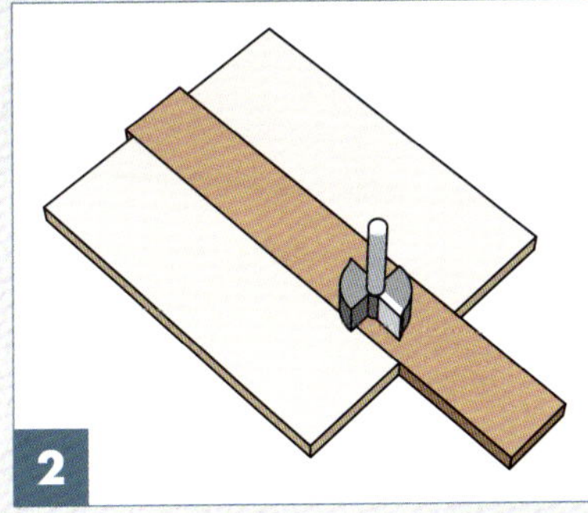

Flatten the front and back surface.

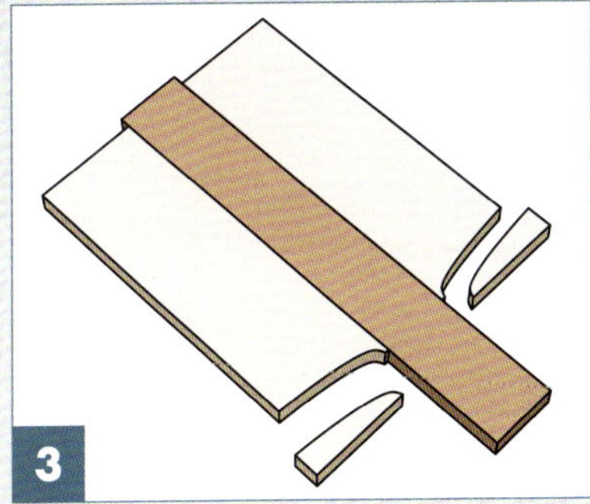

Roughly cut out the curved neck section.

Use a template for the curved work.

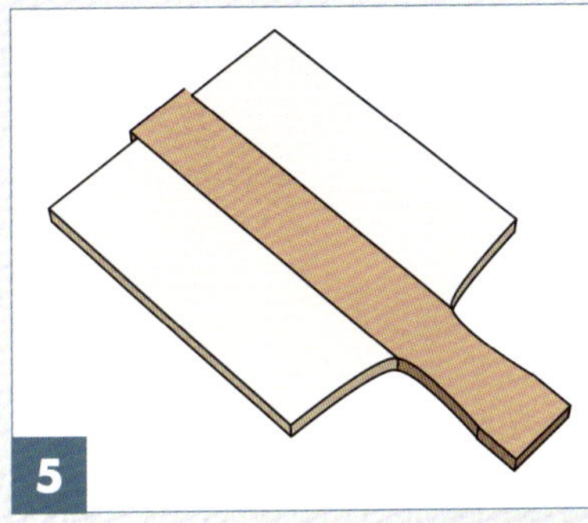

Complete the curved cuts.

Create a template for processing the grooves.

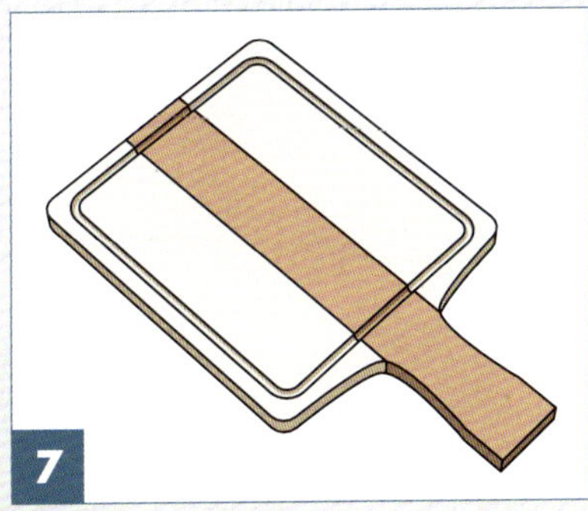

Cut the grooves. Next, round the cutting board curves.

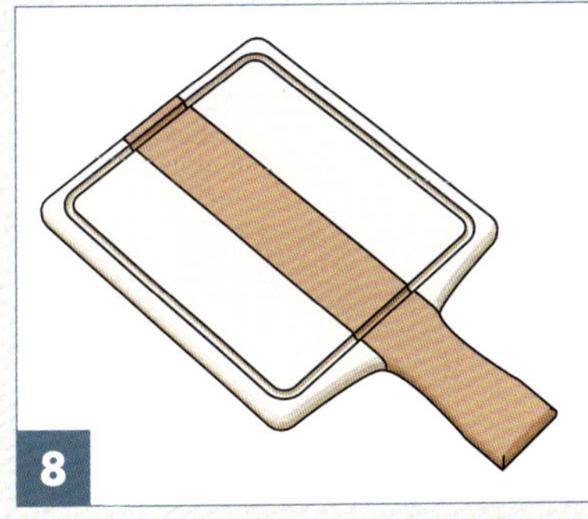

Use a round-over bit to cut the edges.

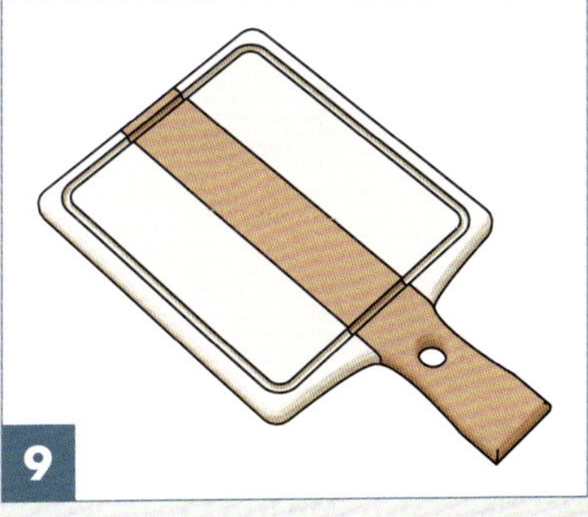

Drill holes and round the edges of the holes.

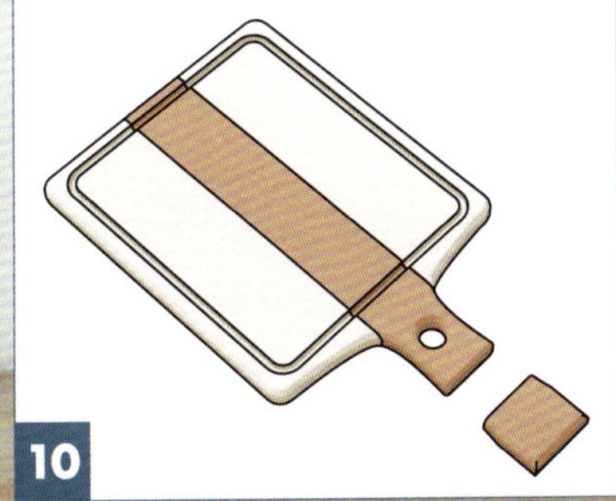

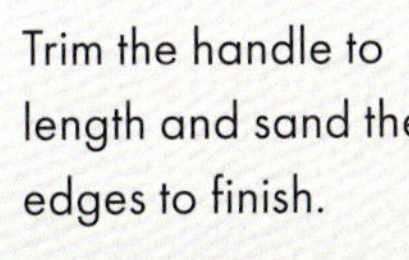
Trim the handle to length and sand the edges to finish.

LAMINATING MATERIAL

1 Cut the material to approximately the desired size and laminate into a single board using DIY clamps or thick rubber bands. There will be a few millimetres of unevenness, but this will be addressed later.

2 Apply glue to the joints.

3 For lamination, ensure that one side is as flat as possible and then apply masking tape. This tape will prevent glue from getting onto the bolts of the DIY clamps that will be used later.

4 While tightening the clamps, continuously check and adjust to ensure there is no unevenness on the joined side. This side will serve as the reference surface (bottom) when flattening both sides.

5 The opposite side may have a few millimetres of unevenness, but this is not a problem. Wait for the glue to cure.

USING A TRIMMER PLANER

6 While waiting for the glue to set, prepare for trimmer processing by removing the auxiliary base from the trimmer and attaching a transparent housing to your DIY carriage.

7 Use a large diameter trimmer bit with a short cutting length to efficiently flatten the large areas.

8 Once the glue has set, remove the clamps. On the flatter side, remove any excess adhesive with a scraper. Further process with sandpaper or a hand plane to ensure there are no protrusions. This surface will be fixed to the workbench.

9 First, fix two rails to the workbench using double-sided tape. The rail height should be about 5⁄64in (2mm) higher than the thickest part of the cutting board. The distance between the two rails should allow the trimmer bit to pass both cutting board edges.

10 Apply double-sided tape to the board. The rail grooves rails in the photo are not significant.

11 Stick the board to the workbench. If there is any wobbling, insert thick paper to stabilize.

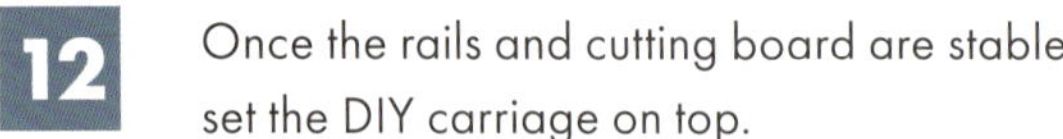

12 Once the rails and cutting board are stable, set the DIY carriage on top.

13 Set the trimmer bit cutting depth. Lower the bit onto the thin area and lift slightly from point of contact. This way, only thick parts will be cut.

FLATTENING PROCESS

14 Move the carriage forward and backward while shifting it incrementally sideways. The aim is to flatten the entire surface. This completes the first pass of the flattening process. There are still slight offsets. For the second pass, set the bit to contact a thinner section and process in the same manner as during the first pass.

15 The entire surface has been cut and flattened. Now, apply double-sided tape to this surface, flip it over, and cut the opposite side.

16 Both sides have been nicely flattened. The surface is still just slightly rough.

17 Use a plane or sandpaper to smooth the surface.

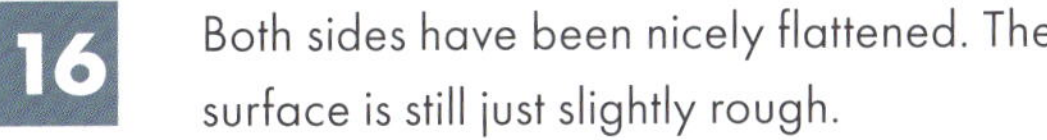

MAKING TEMPLATE

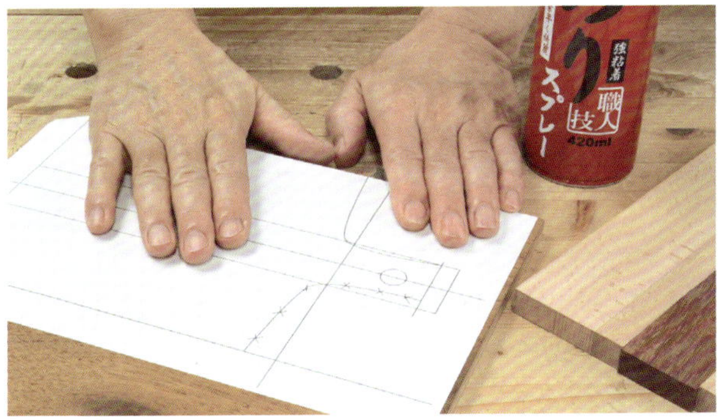

18 Attach a drawing of the curved section (neck) to a ⅜in (9mm) plywood sheet to create a template. The neck shape is basically freeform.

19 Cut the plywood along the curve using a jigsaw or similar tool.

20 Smooth the curved section to finish. Use a drum sander attached to a drill press to more easily smooth curved surfaces.

21 To minimize trimmer bit cutting, it's a good idea to roughly cut away curved portions first.

22 Attach the template to the cutting board using double-sided tape. Only a small portion will be removed by the trimmer bit.

23 The template is placed on the bottom for processing with a flush trim bit. There's a gap between the bearing and blade. Set the bit to a height that will prevent leftover material.

24

For the inner radius, move the trimmer bit from outside the cutting board towards the centre. This allows for a climb cut (refer to photo on right page). You can see that the cut area appears blackened. This occurs when the bit is dull, or the feed rate is too slow. This can be trimmed and corrected later.

25 This is a top-down view. The large DIY base plate ensures stable cutting. To curve the opposite side, attach the template to the reverse side of the board.

PROCESSING WITH A PATTERN BIT

26 Here we introduce a method that uses a 'pattern bit' with a bearing at the base. While the flush trim bit has the template attached below the cutting board, the pattern bit will have it on top.

27 Secure the template with double-sided tape.

28 Set the bit so that the bearing contacts the template. As before, work from the outside towards the centre to achieve a climb cut.

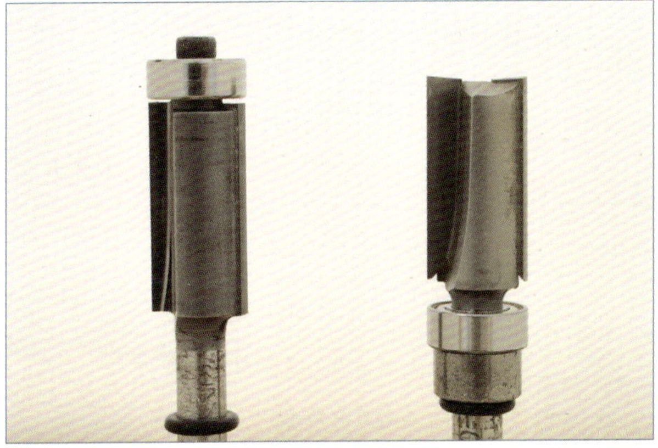

29

There is a difference in naming between bits with bearings at the tip and those with bearings at the base. Both types have the same diameter for the bearing and the cutter, allowing them to create replicas that are the same size as the template. It's convenient to have both types on hand.

From Left: Flush Trim Bit/MLCS Pattern Bit (Whiteside Tools GR)

◉ CLIMB CUTTING VS CONVENTIONAL CUTTING

Correct

Process from outside towards the centre. In the diagram, the cutting surface is not perpendicular to the wood grain but slightly angled. It's important to cut in the correct direction (climb cutting).

Incorrect

Cutting from the centre outwards creates a conventional cut. This can result in a rough surface and make the edges more prone to chipping. To ensure climb cutting, adjust the bits you use or change the surface where you attach the template.

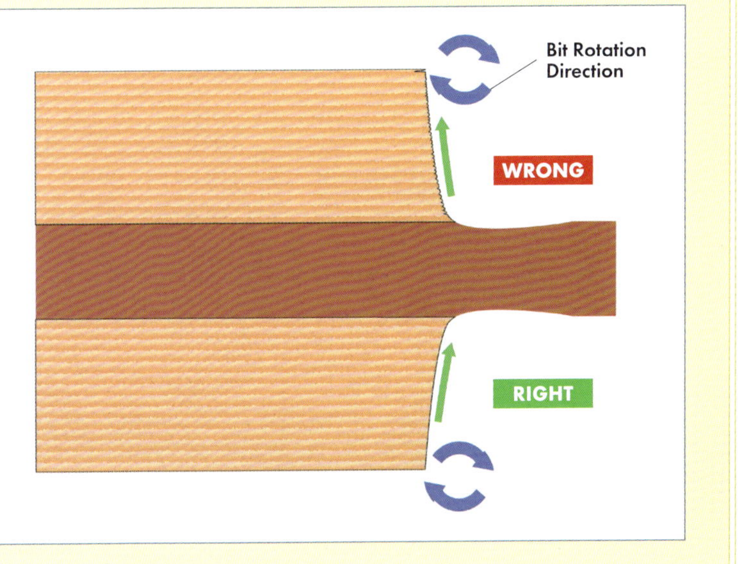

GROOVE PROCESSING WITH TEMPLATE AND TEMPLATE GUIDE

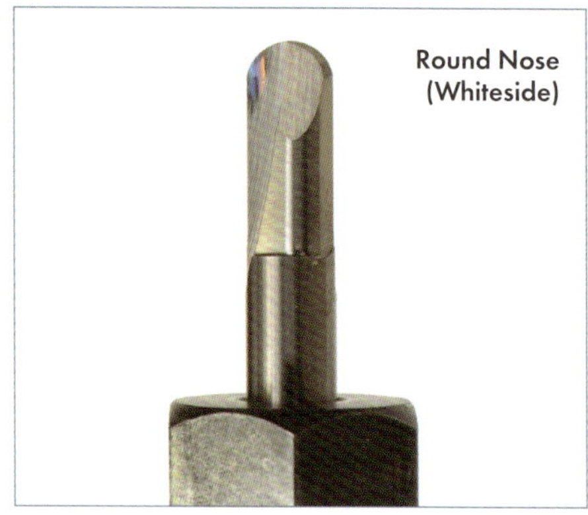

30 This method uses a router bit without a bearing for groove processing. A component called a template guide is attached to the trimmer for this purpose.

31 The bit used here creates a U-shaped bottom groove. Cutter diameter is ¼in (6.35mm).

INSTALLING TEMPLATE GUIDE

32 Remove the sub-base and insert the included template guide into the central round opening.

33 Attach the auxiliary base. The template guide extends slightly beyond the base surface, and the trimmer bit is also exposed.

TWO PROCESSING METHODS

34

The template is affixed to the cutting board with double-sided tape. The surrounding area is grooved. Before making contact with the material, turn on the switch and allow the bit to spin. It's extremely dangerous to switch on the trimmer while in contact with material. Never do this! With this in mind, two processing methods will be introduced. If you feel unsafe, do not attempt this operation, as it is meant to be done using the plunge function on a router.

35

The first method is to release the transparent housing tightening lever, allowing the motor to move up and down. During actual processing, start with the bit above the material. Turn on the switch, and once rotating, lower the motor to begin groove cutting. To prevent the motor from dropping too low and creating a deep groove, pre-set the first cut depth using a hose clamp. If a deeper groove is required, repeat the process. The top circle indicates where the hose clamp is set. The lower circle shows that the bit is slightly lifted above the material.

36

The second method involves setting the cutting depth and starting with the lever closed. Slightly tilt the trimmer so that the bit doesn't touch the material. Turn on the switch, return the trimmer to a horizontal position, and proceed with groove cutting. The template guide must be in contact with the template. Repeat this process to achieve the groove depth.

37

Move the trimmer in the direction of the arrow while making a full circle around the template. Rehearse this in advance and consider your posture as well.

Manufacturer: Justool

38

Finish by rounding edges with a round-over (bullnose) bit. The surface was finished with oil.

◉ ROUND CORNERS OF MATERIAL

You can achieve a clean rounded finish on the templates corners and cutting board by using commercially available templates (pattern boards). These come with four different radii.

Radius Template (Eagle America)

Use a pattern bit or a flush trim bit for processing.

Beautifully rounded corners.

Cutting board corners also rounded in this way.

About The Aluminium Right-Angle Guide

To help you understand the right-angle guide used throughout this book, here is a brief overview. This guide was developed and commercialized as part of a technique called 'Saw Woodworking', which involves minimal use of power tools. Its basic function is to enable you to make straight, right-angle cuts using a handsaw. The magnetic force pulls the saw blade towards the guide. This helps the saw make perfectly straight and square cuts. This is a very useful tool, even for those who regularly use power tools. You can make one yourself (refer to page 189), but if you opt for a commercially available Aluminium Right-Angle Guide, you can attach the included 45-degree pin to achieve highly precise 45-degree mitre joints for projects like photo frames (see page 198).

The saw has a blade without set teeth. Since there are no set teeth, blade thickness will be the exact width of the cut. This will improve the precision of the material's cut dimensions. Additionally, when the saw blade moves back and forth over the magnetic sheet surface, it's less likely to cause scratches. Special blades without set teeth are available for woodworking. These blades offer better straight-cut accuracy when cutting materials at a right angle. However, while using a blade without set teeth allows for highly precise cuts, it tends to get pinched by the material. Depending on the material, you may occasionally need to apply candle wax to the blade or spray it with silicone.

For a detailed description of 'Saw Woodworking', please refer to the two books listed below:

* *Innovative Woodworking Techniques* (Studio Tac Creative)
* *Everything About the Sugita Method of Saw Woodworking* (Ohmsha)

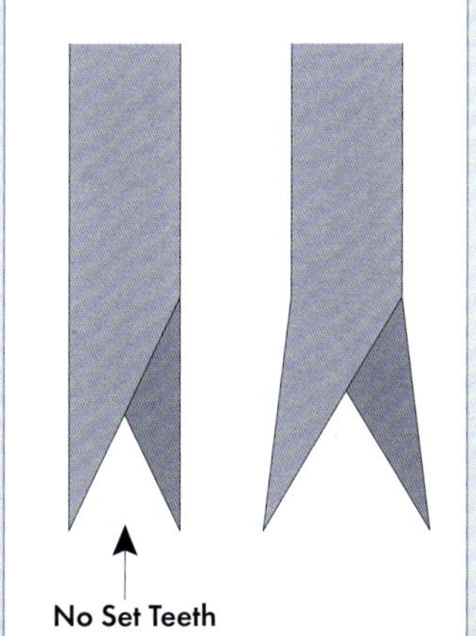

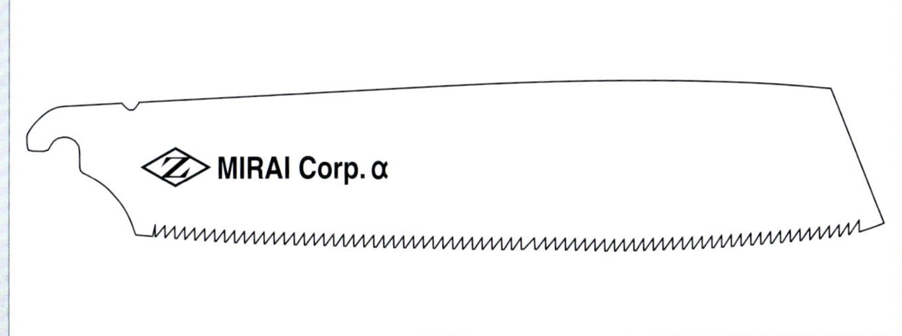

Featherboard 1

A featherboard is a tool used to press material against a fence or router table to ensure precise machining. It is also useful for safely machining narrow or short materials where there is a risk of touching the blade. The featherboard helps keep fingers away from the cutter (trimmer bit).

A featherboard has many cut-outs, each referred to as a 'finger'. Each finger provides a spring effect that pushes against the material. In the photo on the right, you can see a featherboard fixed to the router table and another fixed to the fence. To prevent the table featherboard from moving, there is a backup piece of material in front.

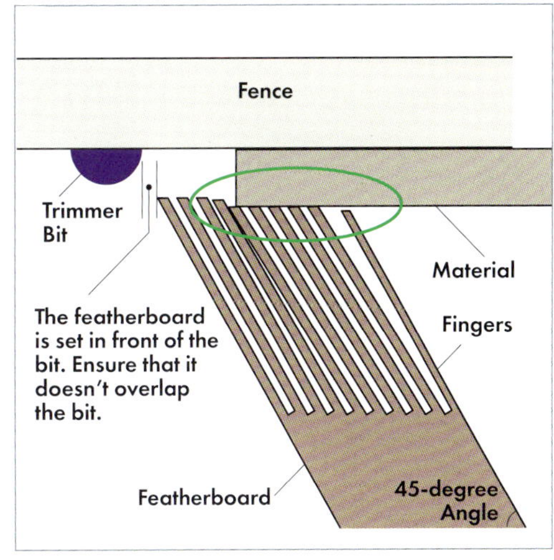

As shown in the photo, the first finger is slightly short. There is no specific angle for the fingers; the author chose 45-degrees, but many commercially available featherboards are around 30-degrees.

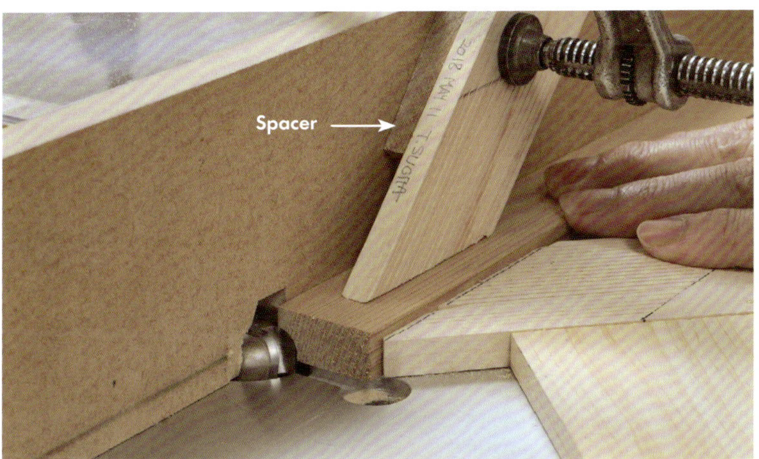

A featherboard, fixed to the fence, presses material against the trimmer table. That being said, pay attention to the contact position. It presses down on the area that is not being cut by the bit, avoiding the portion that will be removed. This is why a spacer is placed between the fence and the featherboard.

When processing thin materials, it's safer to work with a wider piece whenever possible. Then cut that piece down to size with a circular saw or similar tool.

When laminating materials to create frame components, the featherboard is used to safely shape the original thin material.

MAKING A FEATHERBOARD

1 We are just making cuts at regular intervals, so it is easy to just use a bandsaw or circular saw. I, however, used a handsaw.

2 First, fix two pieces of material side by side in a bench vice. Place the jig (see the diagram below) on top. Clamping two pieces together makes it easier to move the jig.

MAKING A CUSTOM JIG

Units: mm

We made a simple jig that uses the help of a magnetic sheet to cut end grain material at right angles. We also created a 'saw blade spacer' by double-sided taping a piece of scrap wood (3⁄32in/2.7mm plywood) to a saw blade of the same thickness as the one we'll be using. Both the saw blade and the saw blade spacer must not have any set teeth.

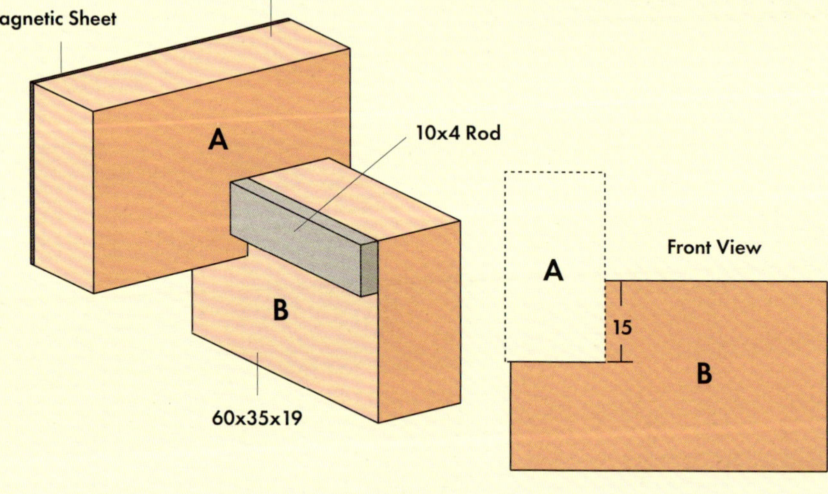

3 Attach the saw to the magnetic sheet and make a vertical cut. Continue cutting along the jig until you reach the pencil line. This completes the first cut.

4 Insert the saw blade spacer into the first cut, align the jig, and determine the position of the next cut.

Make the second cut in the same manner.

6 Repeat the process. Fixing two pieces of material together, side by side, from the beginning makes certain that, as the work progresses, the adjacent piece of material supports its unstable partner as there is less space for the jig on top of the material.

Use the aluminium right-angle guide (see page 233, typically used in handsaw woodworking to cut angled finger-like notches. With this, the process is complete.

Featherboard 2

The featherboard on page 234 has a different shape than a bird's feather. However, it functions the same by pressing material against a fence. The ⁵⁄₆₄in (2mm) boards act as 'fingers', providing a 'pressing' effect. For this featherboard, notches for fingers are cut into a square bar using a handsaw. They can also be processed with a bandsaw or circular saw. Each finger is 4in (100mm) long, and the completed featherboard is 17¾in (450mm) in length.

PRODUCTION STEPS

1

Use an adjustable T-square to set the appropriate angle. Mark with a line. Cut at this position. Spacing between the fingers should be narrower near the bit ⅜–¾in (10–20mm). Wider spacing is acceptable for the remaining fingers.

2 Use an adjustable T-square and place the angled cutting jig on the material. The black surface is a magnetic sheet.

3 Here we see the front and back of the angled cutting jig. Sandpaper is attached to some parts to prevent it from moving when pressed against the material. Size as needed.

4 Make cuts with a handsaw. The stopper is attached to the saw blade with double-sided tape to keep the cutting depth consistent. The blade is 0.4mm thick and doesn't have set teeth.

5 Insert a saw blade of the same thickness as the one being used into the cut groove.

6 Measure the 5/64in (2mm) material that will be used for fingers. Then, subtract the saw blade's thickness from that. Use a spacer of this thickness between the saw blade and jig.

7 Remove the saw blade and spacer without moving the jig.

8 Make another cut.

9 Force the cut-out parts with a screwdriver or chisel. This will create a slit for inserting the 5⁄64in (2mm) finger material.

10

Insert 5⁄64in (2mm) material into the slit. If the material is too tight, adjust by slightly thinning with a hand plane or sandpaper

11

Repeat the same steps to make a featherboard with four to five fingers.

makita

Metric to imperial conversion chart

2mm (5/64in)
3mm (1/8in)
4mm (5/32in)
6mm (1/4in)
7mm (9/32in)
8mm (5/16in)
9mm (11/32in)
10mm (3/8in)
11mm (7/16in)
12mm (1/2in)
13mm (1/2in)
14mm (9/16in)
15mm (9/16in)
16mm (5/8in)
17mm (11/16in)
18mm (23/32in)
19mm (3/4in)
20mm (3/4in)
21mm (13/16in)
22mm (7/8in)
23mm (29/32in)
24mm (15/16in)
25mm (1in)
30mm (1 1/8in)
32mm (1 1/4in)
35mm (1 3/8in)
38mm (1 1/2in)
40mm (1 5/8in)
45mm (1 3/4in)
50mm (2in)
55mm (2 1/8–2 1/4in)
60mm (2 3/8in)
63mm (2 1/2in)
65mm (2 5/8in)
70mm (2 3/4in)
75mm (3in)
80mm (3 1/8in)
85mm (3 1/4in)
90mm (3 1/2in)
93mm (3 2/3in)
95mm (3 3/4in)
100mm (4in)
105mm (4 1/8in)
110mm (4 1/4–4 3/8in)
115mm (4 1/2in)
120mm (4 3/4in)
125mm (5in)
130mm (5 1/8in)
135mm (5 1/4n)
140mm (5 1/2in)
145mm (5 3/4in)
150mm (6in)
155mm (6 1/8in)
160mm (6 1/4in)
165mm (6 1/2in)
170mm (6 3/4in)
178mm (6 7/8in)
180mm (7in)
185mm (7 1/4in)
190mm (7 1/2in)
195mm (7 3/4in)
200mm (8in)
210mm (8 1/4in)
215mm (8 1/2in)
220mm (8 3/4in)
230mm (9in)
235mm (9 1/4in)
240mm (9 1/2in)
250mm (9 3/4in)
255mm (10in)
257mm (10 1/8in)
280mm (11in)
305mm (12in)
330mm (13in)
355mm (14in)
380mm (15in)
405mm (16in)
430mm (17in)
460mm (18in)
485mm (19in)
510mm (20in)
535mm (21in)
560mm (22in)
585mm (23in)
610mm (24in)
635mm (25in)
660mm (26in)
685mm (27in)
710mm (28in)
735mm (29in)
760mm (30in)
785mm (31in)
815mm (32in)
840mm (33in)
865mm (34in)
890mm (35in)
915mm (36in)
940mm (37in)
965mm (38in)
990mm (39in)
1015mm (40in)
1040mm (41in)
1065mm (42in)
1090mm (43in)
1120mm (44in)
1145mm (45in)
1170mm (46in)
1195mm (47in)
1220mm (48in)
1245mm (49in)
1270mm (50in)
1295mm (51in)
1320mm (52in)
1345mm (53in)
1370mm (54in)
1395mm (55in)
1420mm (56in)
1450mm (57in)
1475mm (58in)
1500mm (59in)
1525mm (60in)

Measurements

The measurements in this book are stated in millimetres, which can be converted to inches using the chart above. It is good practice to choose one or the other rather than use a mix of metric and imperial.

Safety

Woodworking is inherently dangerous. Improper use of tools, especially power tools, and disregard for safety, can lead to serious injury or death. The publisher accepts no responsibility for anyone's safety when using this book. Take extra care when working. If you don't feel it's safe for you to perform the techniques described in this book, don't do them until you receive the proper training.

Restrictions

The unique jigs, guides, tools and equipment introduced in this book were invented by the author, Toyohisa Sugita. There is no restriction on readers creating their own similar items and using them for personal enjoyment. However, commercializing or selling them without permission is strictly prohibited.

About The Author

Toyohisa Sugita was born in Tokyo in 1951. At the age of 28, he began building a 27.5ft (8.4m) cruising yacht from scratch. After 5 years and 6 months of difficult but rewarding work, the yacht was completed and launched. This experience marked the beginning of his woodworking journey. He has since been involved in the development, manufacturing and sales of woodworking tools, producing and selling woodworking videos, and has served as a dealer for Lie-Nielsen. His books include The Definitive Guide to Professional Woodworking with Routers and Trimmers and Enjoy Weekend Woodworking (Gakken Publishing). Two of his notable works, Innovative Woodworking Techniques (Studio Tac Creative) and Everything About the Sugita Method of Saw Woodworking (Ohmsha), focus on precision woodworking that can be done with minimal use of power tools.

This book features various jigs. It was through my yacht-building experience that the importance of jigs became apparent. This subsequently inspired the introduction of the numerous jigs found here. A huge number of operations can be performed using a combination of jigs and a trimmer.

Trimmers/routers have always been considered to be highly versatile tools when compared to other power tools. They are capable of drilling holes, making grooves and rabbets, and shaping around templates — all with a single machine. Their versatility stems from the incredible variety of available bits. The more bits you have, the broader the range of operations you can perform.

I believe many of you already own a trimmer and I encourage you to make full use of it!

A trimmer table will also open up a whole new world. I encourage you to use this book as a guide to explore that world.

My previous two books proposed a new style of woodworking that uses minimal power tools, what I call 'saw woodworking'. In these books, I introduced a number of highly convenient processes, such as those for cutting materials to the same size. While writing the manuscript for this book, I often performed material preparation behind the scenes using handsaw woodworking instead of a circular saw. Since these techniques are effective for readers as well, I have included photos and explanations of some of those processes. If you are interested, I encourage you to read the two 'notable works' mentioned above.

Western-style hand planes also frequently appear in this book. I discovered the Western hand plane at a time when I felt stuck in woodworking that was overly reliant on power tools. Despite the common perception that blade sharpening is difficult, with this type of plane, it's remarkably simple. Even beginners can achieve the same results as seasoned professionals. In addition to flattening material, I use the plane for fine adjustments on glue-ups, such as correcting small misalignments or adjusting angles and gaps. The quality of my work has significantly improved because of these planes. Nowadays, I can't imagine woodworking without a Western hand plane. Go ahead and try it yourself, perhaps beginning with tasks such as correcting misalignments?

Toyohisa Sugita

Suppliers

UK

Axminster Tools
www.axminstertools.com

Charnwood
Charnwood.net

Starrett UK
www.starrett.co.uk

Workshop Heaven Fine Tools
www.workshopheaven.com

Yandles
www.yandles.co.uk

Japan

MIRAI, author's website
www.mirai-tokyo.co.jp

Canada

Lee Valley Tools
www.leevalley.com

US

Hida Tool & Hardware Co.
www.hidatool.com

Highland Woodworking
www.highlandwoodworking.com

Lie-Nielsen Toolworks Inc
www.lie-nielsen.com

Pony Jorgensen
www.ponyjorgensen.com

Rockler Woodworking and Hardware
www.rockler.com

The Woodworking Minimalist
www.thewoodworkingminimalist.com

Woodcraft
www.woodcraft.com

IMPORTANT NOTICE

■ **Woodworking inherently involves risks. Improper use of tools and power equipment, as well as ignoring safety precautions, can lead to serious injuries and even death!**

■ **If you do not feel safe using any of the methods introduced in this book, do not proceed with the method! Likewise, if you believe a method is incorrect, use an alternative method. All actions are your own responsibility. Always prioritize safety while working.**

■ **The unique jigs, guides, tools and equipment introduced in this book were devised by the author, Toyohisa Sugita. While readers are encouraged to create and enjoy their own versions for personal use, unauthorized commercialization, manufacturing, or use by businesses is strictly prohibited.**

Index

This book was first designed and published in Japan in 2018 by Graphic-sha Publishing Co., Ltd.
This English edition was published in 2025 by GMC Publications Ltd.
English translation rights arranged with GRAPHIC-SHA PUBLISHING CO., LTD
through Japan UNI Agency, Inc. and LibriSource Inc.
Photos: Takashi Kajiwara, Toyohisa Sugita
3DCG: Toyohisa Sugita Proofreading: ZERO-MEGA CO.,LTD.
Editing: Naoko Yamamoto (Graphic-sha Publishing Co., Ltd.)
Foreign edition Production and management: Takako Motoki,
Yuki Yamaguchi (Graphic-sha Publishing Co., Ltd)

This English edition published 2025 by GMC Publications Ltd,
Castle Place, 166 High Street, Lewes, East Sussex, BN7 1XU
Coordinated by Japan UNI Agency Inc. and LibriSource Inc.

Translator Kevin Wilson
Publisher Jonathan Bailey
Production Director Jim Bulley
Design Manager Robin Shields
Designer Lynne Lanning
Editor Jane Roe

ISBN 978 1 78494 698 2

A catalogue record for this book is available from the British Library.

Colour origination by GMC Reprographics
Printed and bound in China

To order a book, contact:

GMC Publications Ltd
Castle Place, 166 High Street, Lewes,
East Sussex, BN7 1XU, United Kingdom
Tel: +44 (0)1273 488005
www.gmcbooks.com